Mobile Service Innovation and Business Models

Harry Bouwman • Henny De Vos • Timber Haaker
Editors

Mobile Service Innovation and Business Models

 Springer

Dr. Harry Bouwman
Delft University of Technology
Fac. Technology, Policy &
Management
Jaffalaan 5
2628 BX Delft
Netherlands
w.a.g.a.bouwman@tudelft.nl

Dr. Henny De Vos
Dr. ir. Timber Haaker
Telematica Instituut
PO Box 589
7500 AN Enschede
Netherlands
henny.devos@telin.nl
timber.haaker@telin.nl

FREEBAND

This book is sponsored by the Freeband Communication Project Frux. More information: www.freeband.nl

ISBN: 978-3-540-79237-6   e-ISBN: 978-3-540-79238-3

Library of Congress Control Number: 2008926502

© 2008 Springer-Verlag Berlin Heidelberg

Printed on acid-free paper

5 4 3 2 1 0

springer.com

Table of Contents

3   STOF Model: Critical Design Issues and Critical Success Factors 71
H. Bouwman, E. Faber, E. Fielt, T. Haaker, and M. De Reuver

4   The Mobile Context Explored 89
M. De Reuver, H. Bouwman, and T. De Koning

5   The STOF Method 115
H. De Vos and T. Haaker

Freeband Preface

A comprehensive treatise on 'Mobile Service Innovation and Business Models' is possibly one of the best outcomes one could expect from the large and substantial research effort in modern telecommunication, in which the authors have engaged themselves in the last few years. It produces great value by connecting domains that are of crucial importance to the realization of the modern dream of universal access to goods and services, but are hard to relate to each other because of traditional mismatches in disciplinary habits and terminology. This book provides not only an in-depth study of this highly important and highly modern material but also, and as importantly, the tools needed to achieve effective novel designs.

Connecting the notions 'services', 'mobility', 'innovation' and 'business models' requires in the first instance an in-depth, almost philosophical consideration of the meaning and significance of those terms. They may appear originally as vaguely defined, with a very large and largely chaotic literature referring to them. In this book they receive a historical and thoroughly referenced account and are then very precisely defined and put in formal relation. This fundamental work at understanding and defining succeeds so well that it results in very precise UML-like diagrams that structure the topics, relate them to each other and allow for dynamical representations.

The treatise does not end there, of course. After a clear positioning of the central issues, it embarks on a detailed development of each of the main constituents, always put in the general context where business models play a central role in generating service innovation. Acronyms are not shunned of course, the one to describe the context of business models being termed STOF – to distinguish all the important domains that interact with each other: the Service, Technology, Organization and Finance domains. The structure of each of these is then extensively studied and reduced to more primitive concepts, and their relationships are then defined precisely. This way of formalizing is extremely helpful in providing developers with concrete handles for documenting the design of new services that use novel technology while being well-financed and adequately organized.

Designs must not only be conceived, they also have to be evaluated. Often a design cycle will require iterations in which properties of the design are confronted with the originally desired behavior. A specific design methodology is therefore needed, which supports the trek the designer has to make from his specifications to a solution that meets the specs and even has additional desirable qualities (ethical? aesthetical?). The original definition of terms provides the starting point – the 'behavioral specification'– but from there on the system design ladder has to be descended into a closer, more detailed, structure, behavior and dynamics, the evaluation of critical success factors, the creation of the precise business model specification and the determination of its feasibility, e.g., its overall value and its robustness. The crux of the treatise is devoted to this topic.

The book provides a lot of very valuable data: theory and design strategies for (mobile) services based on sound business modeling. But it does not end there. Its results have already been put substantially in practice within the context of a large research program in which the new, 4G telecommunication environment has been researched and developed – the project FRUX in the program FREEBAND, of which the authors were central protagonists. The 4G ideal is 'being connected wherever, whenever, however', meaning 'realizing the new world of seamless communication'. When FREEBAND started, it was not sure that a completely open 4G environment would be feasible, let alone attractive for service development. In the later chapters of the book it is shown how the added flexibility of 4G provides undreamt of new service generation possibilities, e.g., in the context of police protection and mobile payment services.

It is remarkable that the study of connecting services with business models provides the impetus needed to make the new 4G technology of connecting people effective. This could be called a meta-service provision. This book truly provides the service provider with what he needs to design his services attractively and adequately!

Prof. Dr. Ir Patrick Dewilde
Chairman of Freeband
Delft, March 2008

Science Preface

This book contributes significantly to the strongly needed theoretical underpinning of designing viable business models for mobile services. Equally well, it presents some very practical approaches on business model design for electronic and especially mobile service innovations.

The authors develop a conceptual business model framework, their so-called 'STOF model'. Furthermore they provide a business model design method based on this framework, i.e., the 'STOF method', following a 'holistic' view on business models. It may be pointed out that the discussion offers not only new ideas and concepts, but clearly raises the business models to a new level of insight and understanding.

The core contribution is built around four interrelated perspectives, which make up the term STOF: service, technology, organization and finance. Starting from the macro level, the STOF model discussion provides a sound consideration of the relevant design variables in each of these dimensions. The authors then continue to elaborate on some critical design issues, expected to have a major influence on business model viability. Taking into account the existing literature on business models, the combination of providing a partially validated model on the one hand and a method for business model design approach on the other makes this work quite unique. It remains to be seen whether other researchers and industry parties will build on these results by (a) using this structured approach to test hypotheses on business models and (b) analyze and design business models along the practical guidelines. The manifold detailed examples from different industry sectors, which are spread over numerous chapters of the book, may guide the reader to conduct business model analysis and design. This in itself is obviously very useful for researchers and practitioners.

Condensing it into one paragraph, we may state that the STOF model and method can be used to bring research results and prototypes one step further to the market. We sincerely hope that the book receives the attention of a wide audience, which it clearly deserves.

Prof. Dr. J. Felix Hampe
Faculty of Informatics
University of Koblenz-Landau, Germany
Koblenz, March 2008

Industry Preface

Developing new services and bringing them to the market successfully are the great challenges that many companies and organizations face today. Technological developments in information and communication technology act as drivers and enablers of many service innovations. For example, the distribution of content no longer depends entirely on physical carriers but makes use of a variety of broadband networks, both fixed and wireless. In the public sector, electronic government services have been introduced, and in the health care sector much is expected of electronic record sharing and Tele-health. The ultimate success of these new services is, however, not determined by the technological possibilities but by their capability to satisfy needs better than existing alternatives. New services need to create value for their customers and, equally important, allow providers to capture a portion of that value. This means that services require viable business models, i.e., adoption by users and long-term profitability for the providers. Profitability should be understood in a broad sense; it may involve less tangible societal benefits as well.

Designing viable business models is difficult, as customer needs, organizational resources and capabilities, financial arrangements, and technological possibilities have to be taken into account. Especially in the mobile domain, a joint effort by many parties is required to realize a service offering. Moreover, business models and new service offerings of traditional providers are competing with grassroots services such as user generated content.

This book provides a theoretical framework for the description of business models. Furthermore, it provides practitioners with a structured and practical approach to business model design, taking different perspectives into account. Hopefully this will help to increase the chance of successful market introduction of new services.

Prof. Dr. Mark W. de Jong
Managing Director, Telematica Instituut
Enschede, March 2008

Part I Theory

Introduction

H. Bouwman, T. Haaker, and H. De Vos

Western as well as rapidly developing economies depend on innovation in services for their future growth. However, there is little research and knowledge on developing and designing services. Services innovation helps traditional product innovation oriented companies to fulfill market demand. While service companies need to come up with new concepts and approaches. Service innovation is more and more dependent on information technology and digitization of information processes. At the same time we see that in a global networked economy collaboration with other companies and organizations, not only in the information and communication technology domain, but also in supply chains lead to additional complexity in designing new services. Customers are more and more competent in articulating their needs and expressing their demands. They become co-creators of services. Due to capabilities of modern technologies, and the rapid changing needs and demands, service providers have to respond almost instantaneous, and change their service offering. Service innovation is therefore more and more complex.

Although the economic significance of service innovation is high, a sound knowledge base that allows a better understanding of successful service innovation strategies and implementation of associated service systems is lacking. In this book we focus on service innovation, more specifically on service innovation in the mobile domain. Service innovation in our view is directly related to the business models that support these services. The concept of business models is very useful as it considers value creation and value capturing from service development in relation to the design of the service delivery system. Service innovation and development are in our view closely connected to business model design and innovation. In the popular business literature a wide range of business model definitions exists. Common to all these definitions is that a business model describes the 'business logic' of a firm or service, i.e., the way value is created for customers and providers capture value. The concept of value is therefore central in the definition of the business model concept. For a service, new or existing, to be successful, a viable business model is required. A business model can only be viable in the long run if it creates value for users and providers. That is the bottom line.

The aim of this book is to provide a theoretically grounded yet practical approach to designing viable business models for electronic services, including mobile ones. We develop a conceptual business model framework, i.e. the STOF model, and a business model design method based on this framework, i.e., the STOF method. In our framework we take a 'holistic' view on business models. We consider them from four interrelated perspectives, i.e., service, technology, organization and finance perspective. The STOF model not only considers the relevant design variables in each perspective but also elaborates on critical design issues, which are those design issues that can be expected to have a large influence on business model viability.

We focus on business models for mobile services, i.e., all kinds of innovative services that combine technologies and concepts from the area of telecommunication, information technology and consumer electronics. Mobile services typically require cooperation in complex value networks. We therefore take a network view when describing and designing business models, i.e., a multi-actor business model as opposed to the business model of a single firm. Finally, it is relevant to mention that we take a dynamic viewpoint, i.e., we explicitly take the dynamics of business models due to environmental influences into account.

The STOF method may be viewed as a practical implementation of the STOF model for designing business models. It supports the formulation of balanced design choices for the business model elements. It takes the critical design issues into account, as well as the critical success factors for business model viability. The method provides a systematic approach to new service development and underlying business model design. We have reason to believe that our approach is also valuable for services outside the mobile realm.

This book and the models and cases described herein are the fruit of a long-term research effort that started around 2002, although roots of the project can be traced back to the 1990s. The research was mostly carried out within the Freeband and Freeband Impulse programs, the leading Dutch research programs on advanced mobile technologies and applications. Starting out with first descriptions of business model domains, the dynamic STOF model and design method gradually took shape in an iterative fashion. Iterations involved refinements and enhancements based on literature as well as practical experiences with our models and method.

We believe this book provides a structured view on business models and a systematic approach to designing business models. While such approach is no guarantee for success, we think it prevents pursuing the wrong business models already at an early stage. This saves the costs of investing in services and business models that turn out to be less than profitable on the marketplace. This book is therefore useful for professionals, entrepreneurs, academics and students interested in service development and design, (service) marketing, management, strategy and mobile technology in general. It is extremely relevant for both practitioners and academics. The approach we propose will not be totally new to most people in practice, however the systematic nature of our approach is uncommon in practice and will help practitioners to structure and to understand their daily tasks. For academics this book offers a conceptual basis for research, both on mobile services as well as business models.

This book consists of two parts. In Part I we develop our conceptual models and design method. We start in Chap. 1 with introducing and discussing service innovation, and argue that, in order to realize service innovation, it is also necessary to look at the business models of the services in detail. Chapter 2 deals with the conceptualization of our STOF model. We discuss the main theoretical background of the four core perspectives and lay the foundation for the design method. Also the dynamics involved in the success of business models is discussed. In Chap. 3 we make the transition towards the design of business models and introduce and discuss the critical design issues and success factors for business model viability. Chapter 4 explores the mobile context. It discusses the specific characteristics of the mobile services domain and the trends that are relevant for the core business model perspectives. In Chap. 5 we introduce the STOF method. It provides a practical, step-by-step, approach to designing business models, based on the conceptualization of the STOF model in the earlier chapters. Finally, in Chap. 6 we sketch an outlook for future research.

Part II of this book shifts the focus from the conceptualization of our business model framework to the actual application of the STOF model and method in eleven real life cases. Some chapters focus on the validation of the STOF model. Others show how the STOF method can be used to analyze or design business models for (mobile) services in the consumer market, the public sector and in health care. A more detailed description and characterization of the content of Part II can be found in Chap. 6.

This book is the result of the Freeband FRUX project. Freeband Communication is a Dutch national research program that aspires to attain a leading intelligence position in the area of 4G. The future vision of the fourth generation of communication encompasses communication with a maximum of freedom and flexibility for the user. The user plays a key role, has almost unlimited Internet access and is always capable of interacting with anybody, anywhere. Communication infrastructures, applications and services are effective extensions of people's capabilities. Freeband was supported by AIMSYS, Ericsson, Genexis, ISC, KPN, LIONIX, LogicaCMG, Lucent, Philips, TCNL, TMS-I, Waag, Web Integration, WMC, Yucat, Compuware, BiZZdesign and the following knowledge institutes Erasmus University, Roessingh, Telematica Instituut, TNO-ICT, Delft University of Technology, Technical University Eindhoven, Twente University/CTIT and Free University of Amsterdam. Next to FRUX the Freeband Communication program defines a number of related projects:

- AAF: Adaptive Ad-Hoc Freeband Wireless Communications
- AWARENESS: Context Aware Mobile Networks and Services
- A-MUSE: Architectural Modelling Utility for Service Enabling in Freeband
- I-SHARE: Sharing Resources in Virtual Communities for Storage, Communications, and Processing of Multimedia Data
- PNP2008: Personal Networking Pilot 2008
- BBPhotonics: Dynamically Reconfigurable Broadband Photonic Access Networks
- WICOMM: Foundations of Wireless Communication
- B@HOME: Broadband @ HOME
- TUMCAT: Testbed for User experience for Mobile Context Aware applications

In FRUX (Freeband User eXperience) our objective was to advance our understanding of the design and provisioning of *services*, which create meaningful and rewarding *user experiences*. The focus of this book is not on user experience per sé. User experience was the focus of the work of a related research group within the FRUX-project. We started from the assumption that services have to create user experience. Our research group focused on designing and delivering services and how dynamic business networks can create valuable user-experiences, and capture value from it, by offering, configuring and governing adaptive, context-aware mobile service (bundles). In particular, what are effective organizational architectures and financial arrangements for value network cooperation, that contribute to the viable exploitation of services bundles. The projects was designed to develop guidelines and methods for designing dynamic value network configurations and viable service (bundle)s.

In the FRUX project the following organizations did work together, i.e. Delft University of Technology, Ericsson, Free University of Amsterdam Medical Centre, Telematica Instituut, Free University Business School and Computer Science Departments, TNO-Information and Communication technology, VTs Police, Webintegration, and Waag Society. This book is based on research as executed within the Freeband FRUX project, but is also indebted to projects that preceded the FRUX project, i.e. Business Models for Innovative Telematics Applications (BiTa) and Business For Users (B4U). We want to acknowledge the contributions of a number of persons that have been involved in these projects as well as in FRUX. We value the discussions we had, their contributions to case studies, and their theoretical insights. Our special thanks go to Pieter Ballon, Richard Hawkins, Sander Hille, Els van de Kar, Yasin Karamaner, Stefan Klein, Jente Klok, Carleen Maitland, Brigrit Preissl, Wim Timmermans and Uta De Wehn Montalvo. In the FRUX project Victor Aguayo Moreno, Ziv Baida, Freek Ebeling, Amr Eldin, Ralph Feenstra, Ernst-Jan Goedvolk, Henny Gunther, Erik Van Den Ham, Ferial Moelaert, Bert Van Montfoort, Karin Oerlemans, Yat John Pang, Erik Reitsma, Oscar Rietkerk, Evert Schut, Robert Slagter, Jaap Van Der Spek, Maurice Weijgand contributed to the discussions and research that are presented in this book. We also want to thank Hans Akkermans, Christer Carlsson, Jaap Gordijn, Felix Hampe, Christiaan Holland, Ian MacInnes, Alexander Osterwader, Esko Penttinen, Yves Pigneur, Patrick Stähler, Yao-Hua Tan, Paul Timmers, Rene Wagenaar, Pirkko Walden, and Jason Whalley for the discussions we had on mobile services and on business models.

Chapter 1 Service Innovation and Business Models

H. Bouwman and E. Fielt

Western economies are highly dependent on service innovation for their growth and employment. An important driver for economic growth is, therefore, the development of new, innovative services like electronic services, mobile end-user services, new financial or personalized services. Service innovation joins four trends that currently shape the western economies: the growing importance of services, the need for innovation, changes in consumer and business markets, and the advancements in information and communication technology (ICT).

Service innovation is believed to deliver competitive advantage to economies as a whole as well as to individual companies. In this chapter, we introduce and discuss service innovation and argue that, in order to realize service innovation, it is also necessary to look at the business models of the services in detail. Business models help design viable and feasible services by taking into account relevant customer needs and requirements, technical enablers and technological feasibility, organizational resources and capabilities, and financial arrangements. Moreover, it is also possible to incorporate innovation in business models.

This chapter is structured as follows. We begin by looking at service innovation and defining the core concepts, after which we discuss the importance of service innovation, and examine relevant drivers and trends. Next, we address services and service innovation in greater detail, and discuss design approaches that are relevant from a service perspective. In the final part of this chapter we connect service innovation and development to business model design and innovation.

1.1 Services and Innovation: A First Positioning

Before going on to discuss the relevance of service innovation, we want to provide a definition of services and innovation. In economic literature products cover both goods and services. Within marketing, a product is described as anything that can be offered to a market for attention, acquisition, use, or consumption, in order to satisfy a want or need (for example, Kotler, 1988). Traditionally, a distinction is drawn between services and physical products (or goods), by stressing the intangible nature of services. According to Grönroos (2007, p. 52), a service is 'a process consisting of a series of more or less intangible activities that normally, but not necessarily, take place in interactions between the customer and service employees and/or physical resources or goods and/or systems of the service provider, which are provided as solutions to customer problems'. Services are, at least to some extent, produced and consumed at the same time. Customers to a greater or lesser extent participate in the service production process. In other words, producers and consumers create a service together. Because services involve a considerable amount of human activity, they rarely adhere to a predefined process. Services are perceived as the outcome of a process (the service product) as well as the process it self. Generally speaking, products contain elements of both goods and services in varying degrees. The term (total) offering, or total/integrated customer solution, is used to emphasize the fact that a mix of goods and services is required to satisfy the want or need of a customer, for example, a copying machine with a service contract. Although products (i.e. things) and services (i.e. processes) are fundamentally different, they are intimately and symbiotically linked (Shostack, 1984).

The concept of innovation is an important element of the work by Schumpeter (1934), who argued that innovation serves to create wealth through fulfilment of customer needs with five different types of innovation: new products, new methods of production, new sources of supply, exploration of new markets and new ways to organize business. In more recent literature, innovation is related to technological as well as organizational and institutional innovation. In mutual interaction, these forms of innovation constitute the basics of the innovation process, and are conceptualized as systems of innovation (Hekkert, Suurs, Negro, Kuhlmann, & Smits, 2007). Systems of innovation are analyzed at a national level, as technological systems and as sectoral innovation systems. According to the system of innovation approach, innovation takes place in complex environments characterized by dynamic interactions between institutions and organizations that affect the development of innovations. This implies a shift in vision with regard to innovation from a centralized inward-looking, closed approach mainly driven by technical innovations to an open innovation approach (Chesbrough, 2003). Open innovation is characterized by a sharing of knowledge, critical resources and capabilities within and across the boundaries of organizations, and it is enabled by institutions in an open network environment, allowing for the emergence of new technologies, products, services, processes as well as management practices and business models. There is a growing realization that innovation is interdependent in each of these domains: service innovation requires innovation in business models, while product innovation is directly related to service innovation, and process innovation leads to innovations in business models. Consequently, innovation can seldomly be restricted to the product or service offering or the delivery process, but also involves the way organizations collaborate and the supporting information and communication platforms and architectures.

1.2 The Importance of Service Innovation and Services R&D

As far as national economies and individual companies are concerned, service innovation is a prominent issue. Developments at macro and micro-level are interrelated. Off-shoring services and business processes to India, for example, has implications for individual firms as well as for national economies. For a long time national innovation policies have focused exclusively on supporting technological innovation in manufacturing firms, to a large extent ignoring innovation that took place in the services industry. It is increasingly recognized, not only by national governments and service firms, but also by manufacturing firms, that services, service innovation, growth of and employment in services industry are important economical drivers. Manufacturing firms are finding out that a combination of technological innovations and innovation in services can provide a competitive edge, an example of which are services enabling people to update the software of their domestic appliances or online photo albums with a camera. There are a number of trends that indicate why it makes sense to look at service innovation:

- *Services dominate advanced western economies such as the EU and US.* Because services constitute the main growth sector in advanced economies, productivity growth in services is an important element of overall economic growth. In addition, service innovation is currently growing rapidly in most EU countries, (and even faster in the US, *Reneser*, 2006), and services account for a majority of employment and new job creation in western industrialized countries, as well as increasingly in developing countries, specifically in the off-shoring countries.

- *The need for competitiveness in EU services.* As a result of the new European services directive and other actions towards the internal services market, there will be an increase in competitive pressure in the service industries and, most likely, the need for service innovation. The new emerging economies are also shifting their activities towards services.
- *Innovation in services is poorly understood and less 'visible in current statistics'.* Most of the 'official R&D' in services is recorded in computer and related services, telecommunications and R&D services. Service firms are less likely to engage formally in service innovation. Often, investments by service firms in service innovation are not officially recorded because they take place outside R&D or innovation departments, for instance in marketing or in service personalization.

All in all, the sheer size of the services sector in the overall economy, their potential in creating economic growth and welfare (through considerable opportunities for productivity gains) motivates the interest in service innovation. The *European Commission* (2003) has emphasized the relatively low productivity and performance of many services sectors, while O'Mahony and Van Ark (2003) have pointed at the limited use many services in Europe make of ICT. Nevertheless, due to the as yet small but rising share of services in business (technological) innovation expenditures, policy-makers as well as decision makers in the manufac- turing and service industry seriously need to reconsider their innovation policy and strategy, and to focus on service innovation, if only because investments in innovation by private service firms in the US are considerably higher than in Europe.

From a micro-economic point of view, the *Reneser study* (2006) has shown that service firms, i.e. firms whose main focus is on services, like banks and telecommunication operators, are beginning to tackle service innovation more energetically. Nevertheless, the main focal points of service innovation and the way it is organized, budgeted and managed are designed in a diverse way and there is considerable variety among the particular service firms. Based on the Reneser study, we can draw the following conclusions:

- *Service innovation strategy.* Increasing competitiveness and customer needs are important drivers for service innovation. A dedicated long- term service innovation strategy (and hence management) at manage- ment board level is rare. There are few formalized approaches to deriving service innovation strategies. Although open innovation models feature quite prominently in most cases, there is considerable room to improve collaboration within service firms as well as between service firms and research organizations. In most cases, cooperation with regard to service innovation is poorly developed.

- *Service innovation approach.* Although most service firms have some form of structured approach to service innovation, service innovation is less formalized, more dispersed and less explicitly managed and funded. In some service firms there are high levels of technological R&D as well as technology-enabled, mostly ICT-based, innovation, in addition to service delivery and organizational innovation. Formalized, service-only innovation is the exception rather than the rule. In practice, important service innovation activities are hidden behind labels like business development, service improvement, personalization, et cetera, without being recognized as service innovation. Service innovation is often hidden in client-specific solutions.

- *Service management and development methods.* In about half of the Reneser cases more generic formal management methods were used to manage service innovation portfolios, and at project level about half of the firms involved also used more formalized (mostly rudimentary) models for new service development, mainly based on product development tools. However, none of the firms used service-specific design models, methods, or tools, such as service design, service blueprinting or service engineering, or tools like Quality Function Deployment (QFD), Structured Analysis and Design Technique (SADT) or Failure Mode and Effect Analysis (FMEA) in the service domain.

- *Innovation culture and learning.* Creating an innovation-oriented culture that is in sync with service innovation (in firms, industries and society as a whole) is seen as the key to fostering competitiveness successfully. There is huge interest in and potential for cross-firm and cross-industry (lateral) learning, as well as a need for more fundamental research in the service innovation domain.

- *Use of innovation policies and schemes.* Few large service firms are connected to the innovation policy scene (apart from those that themselves conduct extensive technological R&D). Existing innovation schemes are of limited value to most service firms and most of them find it hard or unappealing to gain access to or take part in them. At the same time, nearly all the analyzed companies have no internal management structure to support the systematic acquisition of funded innovation projects, or broadly supported models for collaboration with research institutions that are active in the service innovation domain.

These conclusions from the Reneser study make it clear that much progress can still be made in the service innovation domain, and that service innovation deserves the attention of managers and scientists alike. In the next section we take a closer look at the service innovation drivers and trends at a business level.

1.3 Drivers and Trends for Service Innovation

Because of increasing competition and more demanding customers firms have to innovate their services. Demographic (e.g. ageing population), socio-technical (e.g. market-readiness for new technology) and socio-economic (e.g. income-level, attention to environment and sustainability) trends influence the needs and priorities of consumers. Idenburg (2005) addresses a number of specific consumer trends: individualization, self-chosen collectivism, informalization, cultural diversity, intensivation, and feminization that affect the need for new service concepts, for instance self-service or community based servicing.

At the same time technical developments offer opportunities for service innovation. Every business depends on the exchange of information and the use of information and communication technologies. New techno-logical developments enable the 'blow-up' of the richness/reach trade-off (Evans & Wurster, 1999). Information and communication technologies help distinguish the information world as separate 'marketspace' from the physical marketplace (Rayport & Sviokla, 1994) and make it possible to exploit virtual value chains (Rayport & Sviokla, 1995). Technological developments like the digitization of information, the increased processing capacity of computer chips, miniaturization and increased mobility of devices, the use of sensors and location technologies, increased inter-operability between services, security, and natural interfaces (Bouwman, Van den Hooff, Van de Wijngaert, & Van Dijk, 2005), enable mature architectures and platforms for knowledge sharing, collaboration, and electronic commerce transactions, anywhere, anytime.

In addition to consumer needs, service innovation is to a large extent driven by competitive strategies. There are some risks involved in adopting a strategy that focuses on service innovation. Services, like information, are easy to copy, while they have the highest impact on value creation when they are successful, as is witnessed by the emergence of giants like Amazon or Google. Furthermore, although new product breakthroughs increasingly depend on non-product characteristics, such as complimentary or auxiliary services, core services themselves also require innovation.

Service innovations that cannot be copied have to be based on unique technical features and require unique capabilities and resources available to the firm or the network of firms that provide the service, for instance highly trained employees or specific technologies, such as search engine technologies. System failures in the service innovation domain occur when firms and employees do not have the proper knowledge, skills and competencies, or the network that may provide them with access to the proper intellectual, technical or financial resources and capabilities. Moreover, firms may not be aware of these gaps in capabilities and resources, or they may be unable to identify the actual need customers have for specific services.

Next, the individual user influences the way new services are created and incorporated into their day-to-day routines. Service innovation is to a large extent user-driven, and directed towards providing a specific user experience. Service innovation is an interactive process in which multiple actors, including consumers, play a role. Service innovation is about co-creation, i.e. users providing feedback with regard to existing services and suggesting alternatives, or even developing their own services or content. Intensified interaction with customers will improve the effectiveness of service innovation. The major issue is how to move from personalized services to asset-based services, i.e. services that are reusable and scalable, and that allow for replicative use. However, technology-based services may cause companies to lose touch with customers, which mean they lose an important source of information for service innovation (Matthing, Kristensson, Gustafsson, & Parasuraman, 2006).

With the growing importance of services, service innovation becomes a more important element in the innovation strategy of a firm, which means that more capabilities and resources have to be made available. Some companies, for instance mobile telecom operators, invest in technical R&D as a driver for service innovations. Firms in the financial service industry try to understand opportunities offered by Web 2.0, or internet-enabled social networks like MySpace. Some service firms have technological R&D investment levels comparable to or even above the levels of manufacturing firms (Howells, 2006). In fact, major firms with a manufacturing background, like IBM and Océ, are developing into service solution providers and are among the first to invest in services innovation in a more formalized way (Meiren, 2006). Other service firms benefit from service innovation performed by others, for instance by making use of white labels in the insurance industry.

Service innovations are driven by much more than R&D alone and often need to be combined with new concepts, new ways of interacting with clients and new kinds of service delivery by (networked) organizations.

This requires a more rational approach to services. The development of new services should move from trial and error towards a more systematic approach to design and control. Service firms discover that service design and development is ill-structured and time-consuming, while time pressure is high (short time to market), knowledge regarding service innovation is tacit and hardly formalized, hardly supported by relevant tools, and customer orientation is hard to guarantee (Simons & Bouwman, 2005). To be successful and create value, firms have to develop service innovation methods and tools. Before we can discuss service innovation and design in greater detail we have to take a closer look at service characteristics.

1.4 A More Detailed View on Services

There are four basis characteristics of (consumer) services that are often emphasized in defining services (Grönroos, 1992):

- *Intangibility or non-material.* The acquisition of services does not result in ownership like in the case of physical products, although it results in a right to receive a service. Services are ideas and concepts that are part of a service delivery process itself. Services are non-physical.
- *Inseparability.* Production and consumption take place at the same time. Services, in contrast to physical products, cannot be stored. Significant parts of the service process depend on the interaction between producer and customer, and the information the customer provides. Most of the time customers are present while the service is produced or their presence is mediated by channels like the Internet, e-mail or telephony.
- *Heterogeneity.* Service outcomes and processes are hard to standardize. Quality control and homogenizing services before service delivery is impossible, in contrast to the kind of quality control we find with physical products. Setting quality standards is, however, helpful. Services can vary in quality and breadth, and they may even fail in the presence of the customer. The evaluation of the quality of a service, in terms of outcome and process, depends on the customer's individual and subjective expectations.
- *Perishability.* The service cannot be transferred or resold. If not utilized, the capacity to deliver the service is wasted, for instance in the case of consultant time or movie tickets. The offering itself and the resources needed to deliver the service are not wasted, but have to be made operational in order to deliver the service again.

Although we initially presented services by profiling services versus products, and by stressing the intangible versus the tangible nature, the distinction between products (goods) and services is open to debate. The distinctions between tangible and intangible, homogenous and hetero-geneous, separated and simultaneous production and consumption, non-perishable and perishable, an object and an outcome or process, value created during a production process or value created in the interaction between producer and consumer, and transfer of ownership versus no transfer of ownership, is blurring. Instead of drawing a distinction between goods and services, it makes more sense to see them as the extremes of a *goods-services continuum*. On the one hand the goods are only delivered, while on the other hand only services are being produced (Vargo & Lusch, 2004a, b). The distinction between services and goods is not as strict as we suggested earlier.

Mostly services require physical products for their production or usage. An air transportation service, for example, requires an aircraft. Moreover, services can use physical evidence, for example a physical airline ticket. Customers can have problems with the mental representation of goods, while services, on the other hand, can to a certain extent be sampled before consumption, for instance judging the quality of the food in a restaurant based on the appearance of the restaurant. Goods are more and more branded, and a brand name in itself is intangible in nature. Moreover, many products are only instrumental to a problem that a customer wants to solve as well. Utensils are used for home improvement, cars are meant for transportation, security software – which is intangible in nature – is bought to prevent possible problems, et cetera.

Inseparability is also open for discussion. Many services are provided in the absence of the customer, for instance cleaning services and product maintenance. Moreover, products are customized on the basis of preferences expressed by customers and delivered just in time to reduce or even avoid inventory costs. Products are becoming more heterogeneous. Although standardization is in the interest of producers, consumers want products that are tailored to their preferences. Using information and communication technology, services can be standardized and tailored to customer needs, deepening customer relationships and enabling mass customization. Manufacturing companies gradually shift towards services. The imperish-able nature of goods is also open to debate: fashion becomes outdated, food can rot; product life cycles are becoming shorter and shorter. On the other hand, the resources and capabilities needed to produce services do not necessarily perish, although capacity has to be available.

If we take things one step further, we could argue that everything provides a service. A broader and more inclusive definition describes services as 'the application of specialized competences (knowledge and skills) through deeds, processes, and performances for the benefit of another entity or the entity itself' (Vargo & Lusch, 2004a, b). This definition sees service provisioning as a dominant logic that includes tangible output (goods). Grönroos (2007) refers to the service perspective as a strategic approach by firms based on either a core service or a core product. Grönroos emphasizes that value is created in the value-generating processes of customers and that providing a service means supporting a customer's activities and processes. Customers want solutions to function as services for them. This preference can be offset by a lower price or a technologically more advanced solution.

The basic characteristics of intangibility, inseparability, heterogeneity, and perishability, affect the development and delivery of services with respect to customer participation and service quality and experience. Services must allow for customer participation (Grönroos, 2007). This requires making clear in what way customers are involved in the front office as well as the back office process. The customer, as co-producer of services, is an integral part of the service delivery process, and participates actively in that process. Because users are also co-creators of services, they are also a very important element of the service innovation process. The perception of the service quality and customer satisfaction are both influenced by the service process (i.e. functional quality), as well as by the outcomes of the services process, i.e. the service delivered to the customer (i.e. technical quality) (Grönroos, 2007). Because users are an integral part of the service delivery process, the user experience of a service is an important issue. For example, a day out at Disney's magic kingdom is more likely to be defined by its designers and its visitors as a magical experience than six rides and a burger in a clean park (Clark, Johnston, & Shulver, 2000). According to Pine and Gilmore (1999), user experience plays a decisive role in which suppliers customers prefer. Customers want more than 'just' a product or service: they want an experience that makes a lasting impression. The traditional focus on cognitive evaluation needs to be extended to include service-elicited emotions and experiences (Edvardsson, Gustafsson, & Enquist, 2006).

1.4.1 Types of Services

Services can be characterized in a number of ways. The most important distinction is between core services and support services. A core service is a supplier's main business, whereas a support service is what makes a core service (or product) possible and competitive. Support services have the potential to enhance the user experience of a core service. For instance, the core service of MSN messenger is text-based communication combined with online presence. Support services are, for instance, profile matching, price comparison and emoticon trade. Core products or services are supplemented by 'peripheral', 'auxiliary' or 'hidden' services (e.g. the way questions are answered or information is provided, service recovery procedures, directions for consumption of the core offer, etc.). These are services that the end-user typically does not see (Grönroos, Heinonen, Isoniemi, & Lindholm, 2000; Normann, 2000). 'Auxiliary services' are, therefore, often non-billable, and although they are not primarily what the customer pays for, they have a large impact on customer satisfaction and the effectiveness of the sales cycle (Grönroos, 2000).

Service typologies can be made on the basis of specific characteristics, for instance the degree of labour intensity, i.e. comparing labour costs with capital costs, for instance in the case of auto repair services versus IT services. Services can also be defined based on the level of interaction and customization, for instance service in retail versus services delivered by lawyers, doctors and architects. Another distinction may be related to the recipient of the service, i.e. people, for example health care and enter-tainment services, or objects (things), for example dry cleaning. The service can be continuous, i.e. electric utility or police, or discrete, for instance cell-phones or season tickets. Some services require subscription or membership, i.e. frequent flyer programs or insurance, while others are more informal in nature, i.e. the use of a public highway or pay phone. Services may be available at a single site, i.e. theatre or barber shop, or on multiple sites, i.e. mail delivery. Services may require customers to be mobile, i.e. theatre, bus services, or they may require the service provider to be mobile, for instance taxi, mail delivery. Services may have to cope with peak demands that will cause delays, for instance telephony and electricity, or peak demands that exceed capacity, like movie theatres or transportation. Services can be directed at the consumer or business market, they can be industrial services related to operations and maintenance or they can be information-based. Services can be classified according to domains like transportation, hospitality, government, financial, entertainment, professional services, IT services, industrial services, et cetera, or they can be based on self-service concepts and the use of hard- and software, or involve other persons at the moment of service delivery.

1.4.2 Electronic Services

As we mentioned a number of times earlier, services are more and more enabled by information and communication technology. Since the emergence of telecommunications, data networks, Internet and, most recently, mobile Internet, services are becoming even more virtual. These virtual services, which are provided via the Internet, are referred to as electronic services. Van de Kar (2004) defines an electronic service as 'an activity or series of activities of intangible nature that take place in interaction through an Internet channel between customers and service employees or systems of the service provider, which are provided as solutions to customer problems, add value and create customer satisfaction.' Two common conceptualizations of the technology-mediated nature of electronic services that emerge are electronic services as information services and electronic services as self-service (Rowley, 2006). Hofacker, Goldsmith, Bridges, and Swilley, (2007) discuss three types of electronic services: (1) complements to existing offline services and goods, (2) substitutes for existing offline services, and (3) uniquely new core services. These electronic services have characteristics that are found both in goods and services, and they also have some unique characteristics of their own (see Table 1.1).

The major difference between electronic services and many traditional services is the role people play in the service delivery process. An electronic service is not delivered by humans but by software programs via computer

Table 1.1. Distinguishing between goods, electronic services, and services (Hofacker et al., 2007)

Goods	Electronic services	Services
Tangible	Intangible, but need tangible media	Intangible
Can be inventoried	Can be inventoried	Cannot be inventoried
Separable consumption	Separable consumption	Inseparable consumption
Can be patented	Can be copyrighted, patented	Cannot be patented
Homogeneous	Homogeneous	Heterogeneous
Easy to price	Hard to price	Hard to price
Cannot be copied	Can be copied	Cannot be copied
Cannot be shared	Can be shared	Cannot be shared
Use equals consumption	Use does not equal consumption	Use equals consumption
Based on atoms	Based on bits	Based on atoms

hardware and communication networks. This has major implications for the service characteristics. Electronic services can be accessed anytime and anywhere. Electronic services are information-intensive: digital information plays a key role and is very easy to duplicate and transfer. The role of the customer is also different in the case of electronic services: customers play a more active role via self-service. Electronic services are less personal and use websites, web forms and/or email. No personal relationship between the customer and the company is required. Electronic services can adapt to the customer via (predefined) options for personalization by the customer or customization by the provider. With electronic services, (unexpected) exceptions are not possible, because the rules are set by software and hardware. Electronic services can be consumer services, for instance services delivered by the media industry, but also services as developed by users themselves and are labelled with Web 2.0. But also eHealth, ePayment services, marketplaces, eTravel, distant education and eLearning, et cetera are services that are provided via the Internet or via mobile networks. Mobile services are a specific subset of electronic services. A mobile service is a service that is offered via mobile and wireless networks. This assumes mobility on the part of the user of the services, the devices or applications. We will discuss mobile services in more detail in Chap. 4.

Although electronic services are an example of pure play – complete digitization of the service channel – in practice we see that *multi-channel* approaches are far more common (Simons, 2006). Firms use multiple channels to deliver services, and look for synergy between channels as well as channel coherence. Channels have to cooperate to maximize overall customer value in such a way that the strengths of each channel are used and the various channels complement each other. A seamless and consistent customer experience across the channels will evoke customer trust, which will reinforce the relationship. From a cost savings perspective it is also crucial to strive for synergy effects between the various channels. We adhere to the narrow definition of channel synergy (Power, 2000) which emphasizes reusing assets to minimize costs.

To summarize, electronic services make it possible to provide services anytime and anyplace. Especially, information-intensive services, as distinguished from more labour-intensive and personal types of services, benefit from the emergence of advanced information and communication technology. Self-service supported by the appropriate software and hardware allow customers to deal with services on their own conditions.

1.5 Service Innovation

Information and communication technology has driven service innovation by providing new information and communication services and by enabling innovation in other services. There has been a shift in thinking about service innovation over the years. According to Salter and Tether (2006), this evolution started with neglect, and then moved via assimilation and distinctiveness, to synthesis. For a long time, service innovation has been considered minimal or none-existent. In the past, the focus has especially been on technology-driven innovation in manufacturing, and the impact of (information) technologies on service processes, resulting in what was called the reversed product life cycle (Barras, 1986, 1990). New technologies lead to process innovations that increase the efficiency of the services provided. Next, service quality is increased due to radical process innovation, and finally new services emerge. In time, service innovation began to receive more attention, and the distinctiveness of services as opposite to products was increasingly emphasized in service innovation (Gallouj & Weinstein, 1997). Nowadays, the focus is increasingly on the complexity and multi-dimensionality of modern services and manufacturing, including the bundling into 'solutions' or 'offerings' (Salter & Tether, 2006).

Two approaches have played a central role in service innovation. One has a strong service focus, ignoring technological developments and focusing on service-delivery process, like skills of the workforce and cooperation between departments within the service provider firm. These types of innovations in services are directed at the quality of the service delivery process and at optimizing customer satisfaction. As a result, for a long time service innovation was associated with incremental changes, like stores staying open longer, service quality based on a personal approach, or loyalty programs. In the alternative approach to service innovation, the focus is on the role played by technology, especially on information technology. To a large extent, this approach can be attributed to the increasing importance of information and communication technology, which support services and service innovation. For the first time, information and communication technology, more specifically the Internet, made it possible for service innovations to open up entirely new markets, for instance Netscape, Google, eBay, SAP, Adobe, EasyJet, Starbucks, and Skype. These new technologies made it possible to move away from the labour-intensive, interactive services that were set in a physical environment. Thanks to information and communication technology, services delivery can be asynchronous and does not require the presence of a service delivery staff. It became possible to separate services and to deliver them at a distance. 'Technology has transformed many former inseparable services into services that can be consumed any time or place' (Berry, Shankar, Parish, Cadwallader, & Dotzel, 2006, p. 57). Moreover, ICT adds intelligence to the service delivery process, based on back office applications (e.g. Customer Relation Management – CRM – , tracking and tracing, multi-channel approaches), redefining the client interface by adding online communication and distributions modes, as well as service marketing (e.g. long tail marketing). Information becomes available that may support innovation in specific service functions along the service process. ICT drives both radical and incremental service innovation.

According to Berry et al. (2006), service innovation should take a holistic approach. They discuss nine drivers for service innovation, i.e. a scalable business model, comprehensive customer experience management, investment in employee performance, continuous operational innovation, brand differentiation, an innovation champion, a superior customer benefit, affordability and continuous strategic innovation. Similarly, Den Hertog (2000) discusses four dimensions that are particularly relevant to service innovation: service concept, client interface, service delivery system and technological options (Fig. 1.1). We extended the service delivery system dimension of the original model by not only including human resources but information systems as a resource as well. The characteristics and capabilities of information systems, as enabled by information and communication technology, play a key role in innovation of electronic and mobile services. Often, service innovation involves a combination of the various dimensions; this means the connections between the dimensions (interactions, complementarities) are also important. For example, downloadable ringtones require an electronic communication network for service delivery. Particular service innovations are then characterized by the combination of innovations in one or more of the four dimensions.

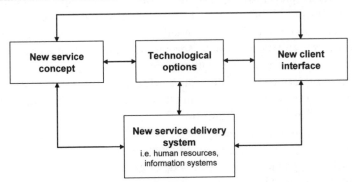

Fig. 1.1. Four dimensional model of service innovation (based on Den Hertog, 2000)

It is clear both from the discussion on innovation in general and from the specific discussion on service innovation that it is important to look at innovation from various perspectives (e.g. customer, service provider, technology) and that a number of disciplines (e.g. marketing, management, finance) have to contribute to understanding and supporting service innovation. We address this when we discuss our business model approach to service innovation (Chaps. 2 and 3), and apply this approach to mobile services. First, we take a closer look at the development and design of new services.

1.6 New Service Development and Design

A new service can be defined as 'an offering not previously available to customers that results from the addition of offerings, radical changes in the service delivery process, or incremental improvements to existing service packages or delivery processes that customers perceive as being new'(Johnson, Menor, Roth, & Chase, 2000). In marketing literature, *New Service Development* usually refers to an overall process of developing new service offerings, while service design refers to the development of blueprints for the service outcome and process. These blueprints can be conceptual (high-level) or operational (low-level) in nature. 'Better service design provides the key to market success, and more important, growth' (Shostack, 1984). Although new service development attracts more and more attention from researchers and practitioners, methodologies and tools that are specific for service development are limited, and depend on product design and engineering. Success factors for new service development are related to the nature of the service, the product-market characteristics, the project synergy, the development process and innovation culture (Johnson et al.). A systematic and formal new service development process is recommended by all studies into the relevant success factors. How such a new service development process should be structured depends on, amongst other things, the type of service innovation and the kind of service (e.g. Fähnrich & Meiren, 2006; Johnson et al.).

Existing literature (Goldstein, Johnston, Duffy, & Rao, 2002; Johnston, 1999; Menor, Tatikonda, & Sampson, 2002; Tax & Stuart, 1997) agrees on the limited contribution made by design methods to service definition and design. Generally speaking, we encounter an engineering-based approach and a marketing-oriented approach. The engineering-based approach starts from traditional product design and then moves on to more specific service design literature, such as service system planning and service blueprinting.

The marketing-oriented approach starts from the service as a process that involves the provider, the user and the quality of the service. The service concept definition, the augmented service offering, and quality function deployment are examples that we will discuss briefly. According to Fähnrich and Meiren (2006), services with a low contact intensity and a low variety may be suitable for the kinds of methods that are used in traditional product development, while the other kind of services require methods that are specifically targeted at services.

Cross (1994) provides an overall framework for describing a product design process. His traditional *Fundamental Engineering* design process covers all aspects of the design process, from problem definition to detailed design. His approach provides a rigorous sequence of steps towards a final result. Service engineering follows the same approach (Luczak, Gill, & Sander, 2006). According to Clausing (1994), technical design principles address only part of the overall design problem, ignoring the customer interaction and service concept. Clausing calls this 'partial design'. According to services literature, partial design and local sub-optimization are among the highest risks in designing and implementing a total service offer and system (Ramaswamy, 1996).

The second method we discuss is *Service System Planning*. Service System Planning adopts a broad approach. The service system is made up of (1) the customer, including needs and expectations, (2) the service concept, (3) the service delivery system, (4) the way the service is perceived by service providers as well as customers, and (5) corporate culture and values, which guides the long term service orientation (Normann, 2000). The design and evaluation of new (additional) services can be aided by looking in turn at the various service system components, and by asking how they will (or should) be affected. Heskett, Sasser, and Schlesinger (1997) provide a lower level insight into the service system, by zooming in on the service delivery system as such. The design of service delivery systems should encompass the roles people play (service providers), technology, physical facilities, equipment and service delivery processes. Assessing these components yields a useful checklist that can prove helpful in the evaluation process by listing the various components and by asking how they will be affected by the new service. However, this method offers no guidelines on how to manage the design process, and nor does it include a rigorous follow-up process that will lead to a finished design. It offers no new design that is based on customer requirements.

The most methodical, process-oriented and design-oriented approach is *Service Blueprinting* (Shostack, 1984). Shostack argues that, compared to the manufacturing systems design, service systems design suffers from a number of problems. Shostack mentions difficulties involved in describing and documenting the processes involved, which lead to intangible results; trial and error approaches that fail to include tests with regard to completeness, rationality and need fulfilment; the absence of a department supervising the design; a gradual approach to quality controls; and a tendency for systems to be described rather than visualized. Where services are concerned, the traditional flowcharting methods that are typically used in service blueprinting are limited and continue to form the core of the analysis. They do not, for example, chart customer involvement in the service provision, and they provide little insight into the organizational structure and its significance in terms of service processes.

In strategic service marketing, Clark et al. (2000) have introduced an approach based on *Service Concept Definition*. Key elements of a service concept are customer value, form and function, customer experience, and customer and supplier outcome. This approach explicitly defines the service concept as a bridge between the 'what' and the 'how' of a new service. The Service Concept Definition is a 'detailed description of the customer needs to be satisfied, how they are to be satisfied, what is to be done for the customer and how this is to be achieved' (Goldstein et al., 2002 p. 123). In this approach, there is also a direct connection between company strategy and customer value. Dividing a service into the 'what' and 'how' makes it possible to identify service elements, to check them against customer requirements or needs, and then to design and deliver those elements. However, it is a rather limited design methodology. Many detailed steps still need to be made before the concept is ready to be implemented. Managing a design process involves more than having a concept. In fact, this approach at best provides 'a direction or point at the horizon for the design outcome' (Simons & Bouwman, 2004, p. 4), and it certainly does not set an agenda for concrete actions.

Grönroos' *Augmented Service Offering* (Grönroos, 2007) uses the service concept as input. According to Grönroos, to develop a service concept, it is necessary to identify the intentions of the organization. The service concept is the starting-point for the development of a basic service package that describes the bundle of services needed to fulfil the needs of customers or target markets. This bundle consists of core services, facilitating (essential) services (and goods) and supporting services (and goods). In addition to the service package, which is targeted primarily at the service outcome, the augmented service offering also addresses the service process. The process consists of three basic elements (from a managerial point of view): accessibility of the service, interaction with service organization and consumer participation. Finally, the service provider has to manage the company's image and communication, including such activities as sales, advertising, sales promotion and communication. However, while this approach offers a more detailed description of the process involved in moving from a more general level towards a more detailed and implementable design, it lacks the methodological support needed to map customer benefits with service activities, and to make and assess service design choices.

Methodological support for design choices is a key element in *Quality Function Deployment* (QFD) (Clausing, 1994; Cristiano, Liker, & White, 2000; Hauser & Clausing, 1988). QFD is a systematic, matrix-based, visual approach to designing quality products and services. It is based on the principle that the quality of a product should be specified as early as possible in the life cycle. Quality requirements are obtained directly from the customers. A list of customer priorities, in words used by the customers, is used as an explicit yardstick throughout the design process. Moreover, possible service functions and solutions are prioritized according to a matrix that is grounded in customer priorities and connected to competitive scores. QFD uses a series of interconnected matrices that establish the quality relationships between higher-level (i.e. product or service level) design activities and their associated lower-level (i.e. sub-process, subsystem etc.) activities. The higher-level matrices can be used in planning the design concept, whereas the lower-level matrices are useful in detailed design and post-implementation monitoring and improvement. The design standards established early on are carried through to later matrices (Herzwurm, Schockert, Dowie, & Breidung, 2002). The use of these matrices enables and stimulates communication between multi-disciplinary development teams.

Service design becomes even more complicated when one considers the possibilities of bundling several, more or less independent services into a service bundle. Chiasson (1999) presents a model for *Service Bundle Design*. Chiasson argues that bundling requires a formal process to structure the economics and strategic value of the bundle and to deliver it to the market. Issues he considers relevant are the strategic intent of the service bundle, in terms of market and product strategy, and its functional objectives. These functional objectives are discussed for channels, marketing, support systems, billing and telecommunication network consideration. Trade-offs in the design of bundles are a key issue for bundles to meet a short time to market, as well as to be profitable and consistent.

Although all of the methods we discussed so far have several shortcomings (in terms of rigor, customer-oriented prioritizing and the evaluation of service alternatives), they also have characteristics that are beneficial to certain aspects of the design process. Fundamental Engineering makes design steps explicit; service system planning provides a useful checklist of the main components involved in the service process;

Service Blueprinting contains a genuine process for developing a service (concept) design, providing a visualization of the service system as an integrated whole, including participants and processes. Service Concept Definition lists the necessary service elements, as well as integrating business strategy, more specifically supplier requirements, with customer needs. The Augmented Service Offering integrates the service as outcome, with the service as process. QFD, finally, emphasizes the need to use a complete set of specifications that are traceable to customer requirements, and that optimize communications within interdisciplinary design teams. If there is one thing all the more marketing-oriented methods have in common, it is that they all emphasize the importance of focusing on the customer throughout the design process and including communication in the service design. None of the formal service development methodologies take the fact that the customer will increasingly be a co-creator of innovative services into account. Moreover, these methods have little or no attention for the technologies that enable the new services as well as the organizational setting and the financial issues at stake. Typically, these issues are discussed in business models.

1.7 From Service Innovation to Business Models

As we discussed above, there is a shift from product oriented innovations towards service innovations, and service innovations are driven by much more than technical R&D alone, specifically with regard to information and communication technology. They often need to be combined with new concepts, new ways of interacting with clients and new service delivery organizations. This is particularly relevant when it comes to the introduction of new electronic and mobile services, where there is also a strong influence of information and communication technology as driver and enabler. Moreover, new service development is rarely a well-structured and systematic process, in which methodologies and tools are used in a coherent and systematic way. Due to the diverse nature of (mobile) service innovation, the lack of coherent methodologies, as illustrated before, and fundamental research that drives and evaluates service innovation, methods and tools, there is still little insight into the critical design issues and success factors involved.

Starting from an open innovation perspective (Chesbrough, 2003), we believe that service innovation is only possible in an open networked environment in which multiple actors collaborate in delivering innovative services, each contributing their own specific resources and capabilities.

However, before this can be accomplished, the underlying business models have to be attractive to all the actors involved. The business model addresses the creation of value via service innovation and the capturing of a portion of that value by mediating between customer needs, organizational resources and capabilities, financial arrangements, and technological possibilities (Chesbrough & Rosenbloom, 2002). Business models are not only relevant for analytical purposes, they also help design viable and feasible services by taking into account relevant customer needs, technological feasibility, required and available resources and capabilities, and suitable financial arrangements. The choice in favour of a more physical product-oriented or service-oriented offering is based on the targeted business model of the provider, in particular the intended customer and network value.

In addition, business models in themselves are a potential form of innovation. An impediment for service innovation is companies being locked into their own business models. 'These companies are reluctant to take risks with their own business by installing new technology, products, services, or distribution channels' (Edvardsson et al., 2006, p. 169). Innovation cannot take place in the product or service offering and delivery only, it also requires an innovative approach in the way organizations collaborate and share resources and capabilities, leveraging existing information and communication platforms and architectures, and in the way value is created for the customers and firms involved (i.e. the underlying business models). A well-known example of business model innovation is the introduction of direct sales by Dell in the computer industry.

In the next chapter we take a closer look at business models. We start from a specific service and discuss the elements that make up the business models. Services and service design are used as a starting point. From the discussion on (service) innovation, it is also clear that multiple perspectives have to be taken into account (e.g. customer, service provider, technology) and that a number of disciplines (e.g. marketing, management, finance) have to contribute to understanding and supporting service innovation. In our approach, we ask fundamental questions regarding the viability and feasibility of mobile service innovation from various perspectives in a structured way. Moreover, our approach makes it possible to integrate different perspectives and disciplines. In addition, it facilitates communication between the various people and organizations involved and enhances their shared understanding of the business model, offering them the possibility to discuss and play with different scenarios before detailed designing and actually implementing a service.

Our business model approach is intended to offer a contribution to the development of service R&D. We believe that service R&D is fundamental and involves the development of new tools, architectures and methods to support service innovation. In our view, service innovation is directed more at the development of new service ideas into functioning concepts, and basically non-technical in nature. Application of our business model approach will support the latter type of innovation.

Chapter 2 Conceptualizing the STOF Model

H. Bouwman, E. Faber, T. Haaker, B. Kijl, and M. De Reuver

Services and services innovation are directly related to and dependent on innovations in business models. In this chapter, we discuss the main theoretical background of the STOF model and its four core components, i.e. services, technology, organizational arrangements and financial issues, which will lay the foundations for the method used in designing business models for (mobile internet) services. We discuss the conceptual background and translate the theoretical insights into more design-oriented issues. Then, we make the transition to a model that helps us understand the dynamics involved in the success of business models, i.e. long-term profitability and market adoption of the service to be designed. In the next chapter we will discuss the Critical Design Issues (CDIs) and Critical Success Factors (CSFs) that are part of the STOF approach.

2.1 Business Model Concept, Typologies and Components

In the 1970s, the concept of business model was used to describe and map business processes and information and communication patterns within companies, for the purpose of building Information Technology (IT)-systems (Konczal, 1975; Stähler, 2001). More recently, business models have been related to market structures and the place individual companies occupy within those structures (Porter, 2001). Sometimes, the concept is used to describe co-ordination mechanisms in economic processes, i.e. markets or hierarchies, or to discuss intermediation or dis-intermediation trends (Hawkins, 2002; Mahadevan, 2000; Tapscott, Lowi, & Ticoll, 2000). In other studies, the implementation of a specific market model, for example the e-shop, e-mall or electronic auction, is discussed in terms of business models (Ajit & Van Heck, 2002; Timmers, 1998). Very often, a single aspect is emphasized, for example in the Business to Consumer (B2C)-model for the retail sector (Lee, 2001; Roussel, Daum, Flint, & Riseley, 2000), or Business to Consumer and Business to Business (B2B) are discussed (Alt & Zimmermann, 2001). Recently, business models have been related to peer-to-peer file sharing services (Hughes, Lang, & Vragov, 2007).

In addition, business models are more and more related to strategic choices companies are making (Hedman & Kalling, 2003; Porter, 2001). Strategies are increasingly translated into business models. Nowadays, many business ventures have a limited interest in formulating strategies. Instead, they formulate business models (Hedman & Kalling). To a large extent, strategies determine the basis of the business case: the concrete operational implementation of business strategy in a business model. The business model is given shape by answering questions with regard to customer needs, the way services are provided, the availability and the way in which the necessary technical, financial and human resources and capabilities are put in place, the way processes are defined, et cetera. Information and communication technology, i.e. Internet and mobile technologies, play an increasingly important role, not only in the organizational processes (back office), but also in the channels that are needed to deliver products and services to their end-users (front office).

As a result, the concept of business model is also closely related to that of business modeling, which describes organizational and/or transactional processes, using Business Process Modeling Notation (BPMN), and object-oriented modeling language, such as Unified Modeling Language (UML), or Integration Definition Function Modeling-family (IDEF). The concept of business models has been established in scientific research within a short period of time and can be considered to be multi-disciplinary in nature. In disciplines including strategic management, information systems, innovation studies, economics, e-business and marketing research has been conducted with regard to business models. The overall result, however, is that, although the business model concept is used broadly, as yet a single coherent description is lacking (Mahadevan, 2000).

2.1.1 Definitions

One of the earliest definitions of business models was proposed by Timmers (1998), who stresses the architectural and technology elements: "a business model is an architecture for the product, service and information flows, including a description of the various business actors and their roles, a description of potential benefits for the various business actors, and a description of the sources of revenues". Based on this definition, a number of alternatives were proposed from various disciplines. One of the most complete definitions was probably proposed by Osterwalder and Pigneur (2002, p. 2): "a business model is nothing else than a description of the value a company offers to one or several segments of customers and the architecture of the firm and its network of partners for creating, marketing and delivering this value and relationship capital, in order to generate profitable and robust revenue streams." At the same time this definition was proposed, the authors of this book were working on projects involving business models, and proposing similar definitions. Bouwman and Van den Ham (2003a) emphasize the relevance of drawing a distinction between the flows of goods, information and money. "A business model provides a description of the roles and relationships of a company, its customers, partners and suppliers, as well as the flows of goods, information and money between these parties and the main benefits for those involved, in particular but not exclusively the customer". Intangible sources also have to be taken into account. We emphasize the design aspects of the business model, and therefore we define business models as "a blueprint for how a network of organizations co-operates in creating and capturing value from technological innovation". This definition follows Chesbrough and Rosenbloom (2002) and is expressed in similar words by Haaker, Faber, and Bouwman (2006, p. 646): "A business model describes the way a company or network of companies aims to make money and create customer value". Reflecting on the various definitions and taking into account our experience with discussing and investigating business models over the years, we would now propose the following definition: *"A business model is a blueprint for a service to be delivered, describing the service definition and the intended value for the target group, the sources of revenue, and providing an architecture for the service delivery, including a description of the resources required, and the organizational and financial arrangements between the involved business actors, including a description of their roles and the division of costs and revenues over the business actors"*. It is clear that the concept of service is of central importance in our definition, and in addition, because in our vision services cannot and will not be delivered by a single organization or company, a number of companies will have to work together to create and deliver the service.

The discussion on business models not only focused on the correct definition of the concept of business model, but also on its components and taxonomies. The attention to classifications was especially dominant in the early discussions about business models, whereas more recently the discussion has focused on the components of business models.

2.1.2 Classifications

The attention to the classification of business models was driven by the emergence of the Internet and the new opportunities for companies to conduct business electronically (Afuah & Tucci, 2001). The use of taxonomies made it became clear in what respect the Internet business models would be different from business as usual. Functional integration as well as the degree of innovation were two key dimensions on which Timmers (1998) based his taxonomy of eleven business models: e-shop, e-procurement, e-auction, e-mall, third-party marketplace, virtual communities, value-chain service provider, value-chain integrators, collaboration platforms, information brokerage, trust and other services. Some classifications are based on developments in the area of technology, others on marketing concepts or product types. In some classifications, elements like value creation and strategy play a role. In other words, the concepts and dimensions in these classifications are different for each author, making a consistent and repeatable analysis and comparison impossible. As a result, a large number of business models are mentioned in various other studies (Afuah & Tucci; Deitel, Deitel, & Nieto, 2001; Deitel, Deitel, & Steinbuhler, 2001; Mahadevan, 2000; Raessens, 2001; Rappa, 2001; Rayport, 1999; Rayport & Jaworksi, 2001; Turban, Lee, King, & Chung, 2000). Some business models turn up in a number of classifications, sometimes in slightly modified or more detailed versions. However, the business models discussed in these taxonomies are according to Weill and Vitale (2001) versions of Atomic business models, like Content Provider, Direct to Customer, Full Service Provider, Intermediary, Shared Infrastructure, Value Net Integrator, Virtual Community and Whole-of-Enterprise/Government models. In our view, most of the business models mentioned in the classifications can be traced back to these eight basic models. Moreover, the classifications are not helpful in themselves, because they do not help us understand the causal mechanisms that can explain the success of specific business models, nor do they offer guidelines for developing those business models. Consequently, it is far more important to analyze what elements constitute a business model.

2.1.3 Components

Hedman and Kalling (2003) rightfully point out that relevant literature regarding business models is dominated by descriptions of "specific" empirically identified business models, and that little attention has been paid to the theoretical sub-constructs of these models. Business models are about the components that have to be discussed, designed, catered and put in place to deliver a service to end-users. The latter is more about ontology: an explicit (and often formalized) specification of a (shared) conceptualization (Borst, 1997; Gruber, 1995) into the components or elements, relationships, vocabulary and semantics of the core object (Pigneur, 2004). The fact that the conceptualization is shared means that common concepts can and have to be communicated between the actors involved.

This leads to the question: what are these basic elements or components of a business model? Alt and Zimmerman (2001) suggest that there are a few elements that commonly turn up in definitions of business models:

- *Mission.* Determining the overall vision, strategic objectives and value proposition, as well as the basic features of a product or service.
- *Structure.* This has to do with the actors and the role they play within a specific business environment (a value chain or web), the specific market segments, customers and products.
- *Process.* The concrete translation of the mission and the structure of the business model into more operational terms.
- *Revenues.* The investments needed in the medium and long term, cost structures and revenues.

Afuah and Tucci (2001) see business models as a system of components (value, revenue sources, price, related activities, implementation, capabilities and sustainability), relationships and interrelated technology, while Mahadevan (2000) emphasizes value creation, revenues and logistics, and Osterwalder and Pigneur (2002) are far more systematic in their approach to the concept of business models. Based on what a company has to offer, who it targets, how this can be realized and how much can be earned, they discuss four basic elements, i.e.:

- Product innovation and the implicit value proposition
- Customer management, including the description of the target customer, channels, customer relations
- Infrastructure management, the capabilities and resources, value configuration, web or network, partnerships
- Financial aspects, the revenue models, cost structure, and profit

Recently, Shafer, Smith, and Linder (2005) have offered an overview based on twelve core publications, and concluded that the core components – the number of components mentioned in literature vary between four and eight, while in all, 24 different items are mentioned (Morris, Schindehutte, & Allen, 2005) – can be summarized in terms of strategic choices, value creation, value networks and capturing of value. The existence of so many different classifications of components, illustrate the lack of a common framework. In our approach, we will therefore focus on customer value, and on the organizational, technical and financial arrangements needed to provide a service that offers value to customers and allows the providers of the services to capture value as well. In our view, business models have to focus on four domains: service, technology, organization and finance, and within these domains different components play a role. We will discuss these four domains (see Fig. 2.1) in greater detail, and also take a closer look at the theoretical and technological concepts that are the basis for our framework, and that will lead the design of business models.

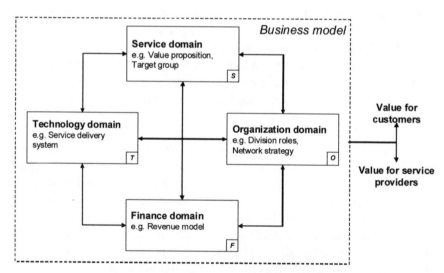

Fig. 2.1. STOF business model domains

2.2 The STOF Business Model Domains

In our opinion, the starting point for any business model is the customer value of a product or service that an individual company or network of companies has to offer and which will satisfy customer demands. We start from the service definition and focus first of all on the value proposition. The service definition serves as a reference for the other domains. The customer value of the service is the most relevant aspect of the service, specifically if one wants to offer a service that really matters to users. Although technology is basically a driver for new innovative services and business models (push-model), from a customer perspective technology is only an enabler. In the latter case technology pull plays a central role, one that can only be understood from a customer value perspective and one that requires an understanding and elaboration of user requirements. After discussing the service and technology domain, we address the organization domain, the way resources are made available, and the finance domain, which includes investments as well as, for instance, pricing strategies.

Each domain description starts with theoretical notions about the relevant concepts and issues with regard to that domain. The concepts that are most relevant from a design perspective are subsequently addressed and included in a descriptive conceptual domain model. The relationships between the concepts within and between the domains are discussed. Pricing, for example, is an important concept in both the service domain and the finance domain, influencing not only customer value, but also the revenues that can be expected. The four descriptive domain models together provide a descriptive conceptual framework for the design of business models. The initial letters of the four domains, Service, Technology, Organization and Finance, together make up our four-letter acronym "STOF", which is the name for our business model framework.

2.2.1 Service Domain

As we discussed before service concepts are the starting point for our approach. There are a number of generic issues that play a role in conceptualizing services, such as customer value. Moreover when we specify the service concept to the mobile domain some other issues ask for closer attention, for example the concept of mobility, and the specific value of mobile services. We will discuss these issues in more detail.

Customer value and innovation. There is a long tradition of literature concerning customer value. Customer value basically consists of what Ansoff's (1987) matrix, based on the dimensions of market and product newness, illustrates. Newness is a fairly troublesome concept, since it may refer to products that are new-to-the-world (Booz et al., 1982), as well as to major (Lovelock, 1984) or disruptive (Christensen, 1997) innovations. Customer value can be seen as a new, innovative offer by a firm to its customers. In general, we will draw the distinction between products or services that are new-to-the-world, and new versions of existing products or services (see also the concepts of versioning as used by Shapiro &

Varian, 1999). Value is seen as part of an equation in which customers in target markets compare the perceived benefits and total costs (or sacrifice) of (obtaining) ownership of a product or service (Chen & Dubinsky, 2003). The value proposition of a firm must be recognized as being better, and as delivering the desired satisfaction of human needs and wants more effectively and efficiently than competitors do. Human needs and wants are what Kotler (1988) describes as the starting point for marketing: "a human need is a state of felt deprivation of some basic satisfaction, wants are desires for specific satisfiers of these deeper needs and demands are wants for specific products that are backed up by an ability and willingness to buy them" (Kotler). In our modern economy, products and services not only have to satisfy needs and wants, they also have to provide valuable customer experience (Bouwman, Staal, and Steinfield, 2001; Magretta, 2002; Pine & Gillmore, 1999). Although "experience" is a relatively intuitive concept, Boswijk, Thijssen, and Peelen (2005) describe ten characteristics of a meaningful experience, some of which are useful for mobile services as well. These characteristics are for instance a heightened concentration and focus, involving all of one's senses. One's sense of time is altered and one is emotionally engaged. The process is unique for the individual, and entails a process of doing and undergoing something. There are elements of playfulness (flow), although one has a feeling of having control of the situation. There is a balance between the challenge and one's own competencies, and there is a clear objective.

With the increasing importance of electronic networks, i.e. the (mobile) Internet, the channels that play a role in offering a product or service play a more central role. That is why Rayport and Sviokla (1994) draw a distinction between content, that what companies are offering, context, the way companies are offering it, and infrastructure, which enables the transaction to occur. These three factors can play a role in defining the newness of the offering. The product or services can be new. The context can be new, for instance location-based mobile services or the channel via which the transaction is executed: mobile, enabled by mobile or wireless infrastructures. Increasing connectivity to a network is crucial to facilitate the transaction process of products and services. The intangible nature of the offering, specifically in the case of services, and the relevance of information to support a product offering, as well as the increased role customers play in the transaction process (e.g. self-service) (McNaughton, Osborne, & Imrie, 2002), become more important and are more indicative of the service-like nature of offerings via electronic networks.

Perceived value versus experienced value. When designing a service, providers have a general idea about what they intend to deliver. However, in most cases, due to all kinds of organizational, technical and operational constraints, the intended customer value is not the value that will be ultimately delivered to the customer. In some cases, the intended value and the value that is actually delivered may be the same, but even in those cases it is not always the value that will be perceived by the customer. In many cases, customer value as perceived by the end-user has little to do with that which that is envisaged in initial business models, depending to a large extent on personal or consumption context (Chen & Dubinsky, 2003). The classical diffusion of innovation concept of reinvention (Rogers, 1995) or the appropriation concept we find in the domestication approach (Silverstone & Haddon, 1996; Silverstone & Hirsch, 1992) play an important role in explaining the difference between the intended value and the value end-users will attribute to a service or a new technology. Users redefine the value and the way they use technologies and/or services in a way that fits their preferences and their behavior.

Value of mobile services. In discussions regarding mobile services, customer value is defined in several ways (see for a more extensive discussion of mobile services, mobility and other relevant aspect of mobile services Chap. 4). Mobile services can provide many functionalities that transcend space (and time), enabling people to study, play, work and shop literally anywhere and anytime they want. Mobile services can break down the barriers to information access, enhance communication and colla-boration, and allow users to perform various tasks without regard to time and location (Wang, 2008). In our opinion, the notion of ubiquity of mobile services, often formulated in terms of services being available "anywhere, anytime", should be more aptly formulated as being available "here and now in this context" (Bouwman, 2004). Anckar and D'Incau (2002) have identified time critical needs, spontaneous needs/decisions, entertainment needs, efficiency needs and mobility-related needs as relevant value sources. Not every mobile service can fulfill one of these generic needs, never mind all of them at the same time. Recent research on the value of mobile services has shown that convenience-focused ideas are well regarded, i.e. services and applications that help the user in solving (daily) practical problems. During leisure time more hedonistic experiences become important, in other words playful and fun aspects are driving value (Bouwman, Carlsson, Molina-Castillo, & Walden, 2008). Social aspects of the technology are important as well, i.e. human interaction and communication, providing a sense of belonging is crucially important (Klemettinen, 2007).

Personalization and context-awareness are important technological enablers that have the potential to enhance the customer value of mobile services. Personalization uses customer preferences, for example stored in personal profiles, to adapt services better to suit users. Context-aware services use the users' context or that of relevant other entities to adapt the behavior of services automatically. With regard to location-aware services (Hegering, Küpper, Linnhof-Popien, & Reiser, 2004). Kaasinen (2005) have found that user needs revolve around five topics, i.e. topical and comprehensive contents, smooth user interaction, personal and user-generated contents, seamless service entities and privacy issues. Tafazolli (2005) presents the Wireless World Research Forum (WWRF) reference model that explicitly links the values and the technological enablers or capabilities. In Fig. 2.2 we combine their model with the distinction made by Bergman, Frissen, and Slaa (1995) between instrumental, and and socio-emotional value.

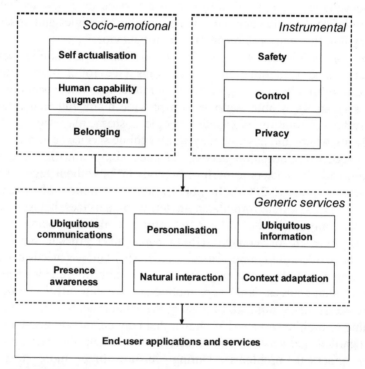

Fig. 2.2. Reference model relating customer value and capabilities

Inhibitors of mobile services development. Carlsson (2006) refers to the mismatch between the mobile services that are being offered and the needs of customers as an explanation for the slow adoption of mobile services. As mentioned earlier, mobile services have to offer a precise match with customer behavior if they are to deliver added value. However, there is a clear trade-off between the customization and personalization of services on the one hand, and security and privacy concerns on the other. Several surveys have identified personal information security and privacy as being among the most pressing concerns when it comes to using new mobile technologies (Haaker, Faber, et al., 2006; Klemettinen, 2007). At the same time, several technologies have been made available to protect people's personal information and privacy. With some notable exceptions, very few of these technologies have been successful in the marketplace. There are several possible explanations for this dilemma. People may not be aware at all of information security risks during certain transactions, and even in situations with full information humans do not always behave rationally. People may also assume that institutions and government organizations are providing a secure platform for their actions (Acquisti & Grossklags, 2004).

Individualization of mobile services. Individualization makes assessing needs and careful targeting crucially important for realizing the potential of providing tailor-made mobile solutions. Market segmentation involves dividing the market (of e.g. consumers) into groups with shared needs or desires, requiring a separate marketing mix (Kotler, Armstrong, Saunders, & Wong, 1996). This means a market segment is a relatively homogeneous collection of prospective buyers. There are many variables on which segmentation may be based, e.g. geographical characteristics, for instance city versus rural area, demographic characteristics, including age, psychographic variables for instance lifestyle, or behavioral characteristics such as purchase behavior or frequency of use (Kotler et al.). Targeting involves the selection of a market segment for which a service is created or at which a service is aimed. Specific market segments may be targeted based on market size, growth potential, fit with the intended value of the service offering, and – as the bottom line – the ability to reach the segment profitably. In the mobile consumer market, segments are often based on lifestyle.

Research in adoption and use of mobile services. Literature regarding the adoption of new technologies is dominated by (1) studies on the Diffusion of Innovation (Rogers, 1995), and (2) studies on the Technology Acceptance Model (TAM) (Davis, 1989), the modifications of TAM as presented under the label Unified Theory of Acceptance and Use of Technology (UTAUT) (Venkatesh, Morris, Davis & Davis, 2003). Many studies have used the TAM concept to investigate the mobile domain (for a recent overview, see López-Nicolás, Molina-Castillo, and Bouwman (draft). Existing Diffusion of Innovation literature suggests that innovations that are perceived by individuals as having greater relative advantage, compatibility, trialability, observability, and less complexity, will be adopted more rapidly (Rogers; Ilie, Van Slyke, Green, & Lou,

2005). The development of routines related to service use increase people's motivation to use new services. Important factors that have a positive effect on adoption and routines are user-friendliness and frequency and regularity of use (Turpeinen, 1998). Consumers evaluate a product/service based on their previous consumption experiences, habits and expectations (Parasuraman, Zeithaml, & Berry, 1988). Past behavior is a strong predictor of future behavior (Schoenbachler & Gordon, 2002).

The TAM is an information systems theory that models how users come to accept and use a technology. The concepts used in TAM are perceived ease-of-use, i.e. "the degree to which a person believes that using a particular system would be free from effort" (related to Roger's complexity concept; Taylor & Todd, 1995), perceived usefulness, i.e. "the degree to which a person believes that using a particular system would enhance his or her job performance", (related to Roger's relative advantage concept; Taylor & Todd), and behavioral intention to use (based on Theory of Reasoned Action (Fishbein & Ajzen, 1975), and Theory of Planned Behavior (Ajzen, 1991)).

Both of these theories on adoption and use of mobile services, however, focus on individual preferences. Another line of thought starts from the use of technology to carry out a specific task or receive a specific benefit within a given context. Especially in the mobile context, with its connotation of "anytime, anyplace", the intention to use a service may depend on contextual parameters as much as on individual preference parameters. Bouwman, Van den Wijngaert, and De Vos (2008) have found that the explanatory value of context parameters is higher than that of individual characteristics when it comes to preferred technology to communicate and exchange information in the context of the Dutch police force. Context plays a more important role than personal characteristics in the preferences for specific communication technology, and could help design artifacts that matter in practice.

Service Design

As we have seen thus far, the central issue in designing a service is "value": a provider intends to deliver and delivers a certain value proposition and customers or end-users expect and perceive a certain value proposition. In our design model, we propose four interrelated concepts: intended and delivered value on the part of the provider, and expected and perceived value on the part of the customer or end-user. We use these concepts to illustrate the match or gaps between the various perspectives regarding "value" (Parasuraman et al., 1988). As we discussed earlier, another important issue in service design is the nature of the innovation.

We propose a distinction between two kinds of innovations: "new version services", which take an existing service one step further (evolutionary), and "way new services", which are new services that are really new (revolutionary) in one or several aspects.

The concepts in service design are defined as followed:

- *Intended Value* is the value a provider intends to offer to customers or end-users of the service. This is the starting point for the innovation, and is often equated with *Value Proposition*, although there often is a gap between *Intended Value* and *Perceived Value* – and we would like to model these gaps. *Intended Value* is translated into functional requirements (technology design), like technical specifications, and into requirements for the value network (organization design), like roles that are necessary.

- *Delivered Value* is the value a provider actually delivers to customers or end-users of the service. Functional requirements are translated into technological functionalities (technology design), which in turn determine the *Delivered Value* – these translations are not straightforward, and there are often gaps and mismatches. *Delivered Value* is also determined by (non-technical) value activities (organization design), like help desk, support and (physical) distribution.

- *Expected Value* is the value a customer or end-users expects from the service, based on their experience with *Previous Versions* (technology design) of the service (in the case of a "new version service"), or with similar services (in the case of a "way new service"). For *Previous Versions*, but also for similar services, it is possible to describe the "backward" and "forward" (technical) compatibility with previous and next versions or generations of services. Furthermore, *Expected Value* is determined by resources and capabilities (organization design), like trust and reputation, and by financial arrangements (financial design) like paying for the device, paying per usage or paying a flat fee, subsidized handsets or discounts.

- *Perceived Value* is the value a customer or end-user actually perceives when they consume or use the service. This is the "bottom line" – it is the customer or end-user who evaluates the value of the innovation. *Perceived Value* is like the difference between *Delivered Value* and *Expected Value*, including functional, emotional and process-related aspects. The higher the *Delivered Value* is, or the lower *Expected Value*, the higher *Perceived Value* will be. There are other variables that also have an important influence on *Perceived Value*: *Customer or end-user, Context, Effort* and *Tariff*. These concepts are defined below.

- *Customer or End-user.* We use the term "customer" to refer to the person(s) paying for the service, and "end-user" to refer to the person(s) actually using the service. In the case of consumer services, these roles coincide, but with regard to business services, they may be separate entities: e.g. the "customer" role can be played by a "decision making unit", and there may be different "end-users", e.g. employees use basic functions, while managers use basic functions plus some additional managerial functions. Within consumer and business markets we can distinguish *Market Segments*, each with different needs, wishes and preferences, which determine *Perceived* and *Expected Value*. Important qualities of the customer or end-user *Market Segment* are: size of "installed base" (already have or use similar services or earlier versions), size of target group, size of maximum potential market. We can also describe percentages of ownership, adoption (usage), access to the service (e.g. not owning it, using it sporadic), or (cognitive or otherwise) expertise levels within specific market segments. Knowing and understanding customers and end-users is crucial to successful innovation – and must serve as a starting point for formulating *Intended Value*.
- *Context.* A service is always consumed or used within a specific context, and an innovation is only successful if it offers benefits in that specific context. The concept of *Context* can be described at various levels of abstraction: there is the context of one concrete situation, e.g. walking on the street or sitting at home, the context of daily life, e.g. working or professional life, or private or family, and there is the wider social-cultural context, including societal trends and political drivers and constraints. Other products or services with similar functions are part of *Context* – the frame of reference in which *Perceived Value* is determined. Note that context can thus relate both to static personal attributes such as professional life, and dynamic attributes such as current location Knowing and understanding *Context* is crucial to successful innovation – and must serve as a starting point for formulating *Intended Value*.
- *Tariff and Effort.* Customers pay a *Tariff* (price) to consume the service, and an end-users make an *Effort* to use the service, whereby *Effort* refers to all non-financial efforts the end-user must make. These two variables have a clear influence on *Perceived Value* and therefore on the adoption and use, and thus on the success of a service.
- *Bundling* of services, or of services and products, is common practice. In Chap. 12 we specifically deal with this issue. Bundling of services in general leads to increased value of services to the customer or end-user.

To estimate realistic levels for *Tariff* or *Efforts*, one may study the existing tariffs or efforts regarding similar services, e.g. a first estimate of a tariff for a news service may be the price of a single newspaper. The effort concept can best be explored by making use of Rogers Diffusion of Innovation concepts. According to Rogers (1995), there are five factors that influence the adoption and use of an innovation: compatibility with people's daily contexts and the relative advantage of the innovation over other products or services (both covered in concepts discussed above), complexity (how easy the service is understood by the customer), visibility of the advantages of the innovation, and testability of the innovation. The last three factors are related to the concept of *Effort*. Ease-of-use is the flip-side of effort. The easier to use a service the less effort one has to do. The introduced concepts in the service domain and their relations are mapped in Fig. 2.3.

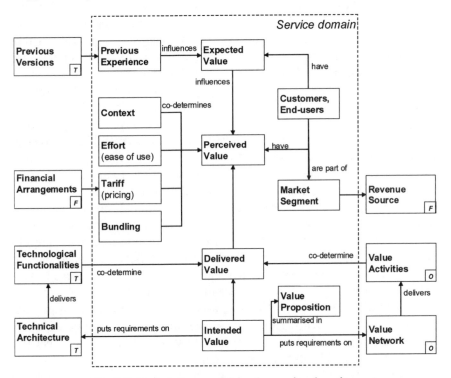

Fig. 2.3. Descriptive model for the service domain

2.2.2 Technology Domain

Requirements as defined in the service domain determine and specify the technical architecture, which is part of the technology domain (Fig. 2.4). In the technical architecture, (mobile) middleware, including web services play an important role, in addition to network and infrastructure characteristics, in facilitating the process that enables the service development, creation, discovery, delivery, bundling, control and management. Business processes can be embedded in web-services. Web services contain both IT-functionality and data.

Organizations have a choice in the degree to which they want to embed processes in IT-functionalities. The most detailed level at which business processes can be embedded is the CRUD-matrix level (Create, Read, Update and Delete). At a higher level, objects are defined. Objects are related to the business and information processes. A complex organization can use object-models with thousands of objects with a limited scope. At still higher-level components are used. Components are applications that can be used by multiple users. At a higher level still, web-services are discussed (Koushik & Joodi, 2000). Functions and objects are combined with business processes in a service application that can be used by "business messages". Web services have the highest level of granularity. They are business functions that are exposed to the web through a well-defined interface, and use standard web protocols like Universal Description, Discovery and Integration (UDDI), Simple Object Access Protocol (SOAP) and Web Services Description Language (WSDL) (Lankhorst, Van der Stappen, & Jansen, 2001). Most web services are based on a 3-tier infrastructure, defining external client interfaces, middleware and application services and back end data services. Web-services can be provided by third parties and do not depend on the IT-resources of an individual firm. Furthermore, we have to realize that, to provide services over the Internet and mobile networks, organization legacy IT-systems or web-services are not sufficient. Mobile web services, which are also discussed in Chap. 4, offer many opportunities to provide generic service elements. Here, we discuss some of the generic technical issues that have to be developed in any service and application, that run over a network and is offered to customers.

Authentication. Authenticating users in a secure way is required to identify the user, e.g. when offering personalized services. In addition, if users are billed for consuming network resources and/or content the user must be identifiable for service providers. Given risks of identity theft and fraudulent behavior, security is a key issue in any authentication process. Generally, diverse solutions exist for authentication, including username/password or device based authentication. In the mobile domain, the authentication of mobile devices has been supported by operators using the Subscriber Information Module (SIM) card model. SIM cards provide user information to the network and a secure link between the subscriber account and the operator bill, and allow for personalization and localization of services (Microsoft, 2003).

Management of user profiles. User profiles containing information such as preferences, personal data, interest and context need to be gathered, stored, and maintained. This is especially relevant when offering personalized services. However, user profiling also can guide service providers' future strategies as they provide feedback on user behavior from which user preferences can be inferred. There are several approaches to storing user information, involving active user involvement to a greater or lesser degree (Pashtan, 2005). Sometimes, user information is distributed among various companies. Although potentially highly interesting, it is not always legally possible to share the information (cross-domain profiling) (Ali Eldin, 2006). Another choice is between automatic user profile generation versus active user involvement. Moreover, user information can be stored decentralized on the device and can be requested by any service provider. The role of user profiling is strongly related to having customer ownership and authentication, as profiling is only possible after identifying and authenticating the end-user.

Security. Success with regard to the demand of a service offering to a large extent depends on the level of security perceived by consumers. Wireless networks have security issues with respect to over-the-air transmissions and additional gateways between wireless and wired domains. Standardization organizations, such as 3rd Generation Partnership Project (3GPP) and Internet Engineering Task Force (IETF), deal with those security concerns.

However, the original World Wide Web Consortium (W3C) specifications for web service standards provide little or no security, which means that external mechanisms or standard modifications have to be arranged, and the question is how to implement these security requirements into Mobile Web Services (MWS) technology (Pashtan, 2005). Association between mobile device and user ID has been traditionally implemented in the SIM card model, which is not a very sophisticated system. This means that a design choice has to be made between security in web services technology or more generic network-based security.

Technology Design

The service design and the generic issues we discussed above, more or less serve as a guide to the technical design. The most important technology design variables, and some of their relevant characteristics are:

- The *Technical Architecture* describes the overall architecture of the components listed below. Important characteristics of the technical architecture are: centralized vs. distributed, open vs. closed, interoperable vs. non-interoperable.
- The *Backbone Infrastructure* refers to the medium and long range backbone network infrastructure. Important characteristics are: high vs. very high bandwidth, future-proof vs. non-future-proof.
- *Access Networks* refer to the first and second mile network infrastructure. Important characteristics are: fixed vs. wireless, high vs. low bandwidth, universally available vs. deployed in hotspots, scalable vs. non-scalable.
- *Service Platforms* refers to the middleware platforms enabling different functions, including *Billing, Customer Data Management*. These platforms provide the generic business functions as authentication, billing and customer care. Specific platforms offer for instance location or context information. Important characteristics are: centralized vs. distributed, personalized vs. non-personalized, secure vs. non-secure, legacy vs. new, open vs. closed.
- *Devices* refer to the end-user devices providing access to services. Important characteristics are: multi-purpose vs. single purpose, "network intelligent" vs. "dumb interface", storage facilities vs. no storage facilities, embedded software vs. open terminal.
- *Applications* refer to the user applications running on the technological system. Important characteristics are: communication vs. content, always on vs. time-critical, personalized vs. non-personalized, secure vs. non-secure.
- *Data* refers to the data streams transferred over networks. important characteristics are: asynchronous vs. real-time, high volume vs. low volume.
- *Technical Functionality* refers to the functionality offered by the technological system. Important characteristics are: always on vs. time-critical, personalized vs. non-personalized, secure vs. non-secure.

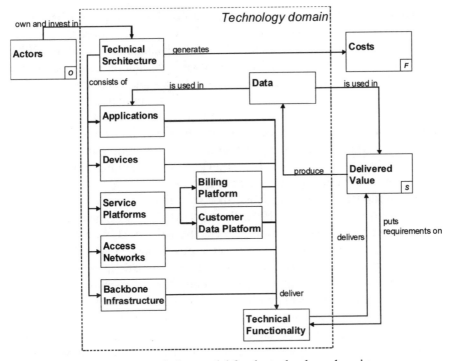

Fig. 2.4. Descriptive model for the technology domain

There is a direct link between the technologies that have to be implemented and used in order to be able to provide the service, and the organizations involved in supplying these technologies. To a certain degree, the technology design is connected to the institutional design: the arrangements between actors regulating their relationships, tasks, response-bilities, allocation of costs, benefits and risks (Koppenjan & Groenewegen, 2005). We discuss such organizational issues in the following section. It may be clear that, depending on the choice that are made, the organizations involved in delivering the resources and capabilities, in terms of technology, finance, marketing and management, may vary.

2.2.3 Organization Domain

In general, organizational issues revolve around the resources and capabilities, mainly related to technology, marketing and finance that have to be made available to enable the service. Although the service may be offered by a single organization we assume that this organization has to collaborate with others in order to be able to provide all the necessary resources and capabilities that are required for developing and offering the service to the market, and to develop a viable business model for involved actors. In their analysis of business models, Hedman and Kalling (2003) conclude that the bottom line is that economic value is determined by a firm's ability to trade and absorb ICT-resources and capabilities, to align (and embed) them with other resources and capabilities, to diffuse them in activities and manage the activities in a such way as to creates a proposition at uniquely low costs or with unique qualities in relation to the industry in which the company is operating. Collaboration, in-sourcing and network formation are possible strategies for obtaining the necessary resources an organization does not have.

In traditional strategic analysis, like specifically theories on Industrial Organization (Porter, 1985, 1990; Porter & Millar, 1985), and the strategy process perspective (Henderson & Venkatraman, 1993; Mintzberg, 1983; Scott Morton, 1990), strategic, economic positioning in an industry sector plays an important role. These approaches are atomistic in nature, focusing on autonomous firms that create competitive advantage in specific markets. Starting from strategy theory, these authors conclude that strategy has to deal with industry, industry position, customer segment, geographical markets, product range, structure, culture, and position in value chain. Value chain analyses have gained popularity through the writings of Porter (1985), and have since evolved to include a wide variety of models. Although the original purpose of a value chain was to identify the fundamental value-creating processes involved in creating a product or service within a firm, the concept has since been broadened and is often used to describe an entire industry. An industry-level value chain serves as a model of the industry, whereby the processes are considered independent of the firms that may or may not engage in them. This separation makes it possible to analyze the positions of various firms in the overall industry, as well as instances of vertical integration or cooperative agreements (alliances, joint ventures, et cetera).

Despite the strengths of the value chain approach, it has been criticized from two perspectives. To begin with, its external focus has been criticized by Barney (1991). In his resource-based view, Barney is more internally focuses, i.e. on the resources and capabilities a firm controls and that enable that firm to offer a compelling strategic advantage compared to others. The core assumption of the Resource-based View of the firm is that competitive advantage of a firm is derived from the resources and capabilities the firm controls that are valuable, rare, imperfectly imitable and not substitutable (Barney, 1991, 2001). A resource can be defined very broadly to include almost anything in an organizational or inter-organizational setting. This is clearly visible in some of the most widely used definitions of the term resource. Firm resources include all assets, capabilities, organizational processes, firm attributes, information, knowledge, et cetera, controlled by a firm that enable it to conceive of and implement strategies that improve its efficiency and effectiveness (Barney, 1991). In literature, little attention is paid to the kind of resources that should be shared in value webs and how they are organized. Although there are several resource typologies (Grant, 1991, Allee, 2000a, b: tangible-intangible or knowledge based resources; Barney, 1991: physical, human and organizational capital resources; Das & Teng, 1998: financial, technological, physical and managerial, Miller & Shamsie, 1996: property-based and knowledge-based). These typologies are too general. We will come back to this after we have discussed the value web concept in more detail.

Secondly, the value chain metaphor has been criticized by more network-oriented approaches. Tapscott et al. (2000) note that the chain metaphor masks the importance of horizontal aspects of a firm's processes, in particular their relationships with other firms, for instance in making resources available. As early as 1978, Pfeffer and Salancik introduced the resource-dependency perspective. Resource Dependency postulates that the power an organization wields within the value network depends on the resource dependency relationship is has with other organizations. This relationship operates in two ways. If an organization can easily obtain a resource, the dependence and with it the power of the organization providing the resource will be low. If a focal organization depends on an important resource from another organization, there is a dependency relationship, and the other organization will wield power over the focal organization. In a relationship each firm can, to a certain degree, withhold or increase the resources in question. Networks are formed based on this notion of the sharing of resources between organizations. A firm will try to form alliances with other firms to ensure a consistent supply of critical resources and in doing so create interdependence between the partners and gain access to the other firm's network. This process allows the creation of a value network that is critical to its survival, as it makes it possible to deal with the uncertainty of the competitive environment, provide flexibility and reduce the need for (risky) investments (Sakaguchi, Nicovich, & Dibrell, 2004).

The resource-dependency view, and the criticism regarding the traditional value chain that dynamic forces in the course of production are ignored and that the value chain model implies that product and service development is necessarily a sequential process, did lead to the development of alternative conceptualizations of the value chain in the form of business webs and value nets (see respectively Kothandaraman & Wilson, 2001; Tapscott et al., 2000). There is a shift in conceptualization that emphasizes the provisioning of information, products and services by networks that consist of sub-units of organizations and/or cooperating organizations working together (e.g. Stähler, 2001). Moreover, the boundaries of organizations are becoming more transparent, and organizations cooperate in changing constellations, enabled by ICT. Information, services, and products can be provided by sub-units within organizations, by single organizations or by collaborations between companies. More and more vertical integrated organizations are redesigned in network organizations in order to be more flexible and to be able to offer new services more rapidly to the market (Versteeg & Bouwman, 2006). We will call this phenomenon, where organizations or organizational units work together to offer new services, value networks, value net or value web.

The broadening of the value chain concept to that of a value net or web coincides with the general trend in which greater attention is being paid to network concepts in strategic management literature (Gulati, Nohria, & Zaheer, 2000). By definition, plural organizations with various roles and functions create an organizational network by pursuing a shared set of objectives (Demkes, 1999). Interorganizational networks, relationships between firms that extend beyond the dyad or triad, come in many forms, such as business groups (Granovetter, 1994), cooperative and governance networks (Wigand, Picot, & Reichswald, 1997), constellations (Jones, Hesterly, & Borgatti, 1997), network enterprises (Castells, 1996), and strategic networks (Gulati et al.). These various forms can be differentiated based on the interaction patterns among the members, as well as the flows of resources between them (Jones et al.). A more dynamic approach, specifically directed towards the evolution of networks, seen as complex systems, is discussed by Monge and Contractor (2003).

With regard to complex value systems, the generation and delivery of value to the users becomes a mutual interest. Based on the resources and capabilities that are available in a complex value system, organizations adjust their functional contribution in the development of customer value. Their operation in this framework is based on the exchange of information, products, services and financial assets. Hence, organizations become dependent on each other, strategically, functionally and financially. Continuous and repetitive interaction leads to the emergence of relationships between firms, which may become institutionalized through legal agreements and contracts. The relationships between the actors can exist at various levels, e.g. communications, information flows and revenue flows (Maitland, Van de Kar, When de Montalvo, & Bouwman, 2003).

Complex value systems or value nets have to strive to support customer processes to the maximum extent possible when they want to improve customer value (based on Grönroos, 1994). On the other hand, there are costs associated with every service. Companies can decide whether they support the entire customer process, or only one or a few stages in the entire process. Their decision depends on what their core competencies are (based on their resources and capabilities) (Petrovic & Kittl, 2003). Since we assume that many services are provided by value networks, this decision is the basis for the configuration of value webs. Each company in the web will choose which value (and ultimately which part of the end-user value) it will offer, or in other words, which part(s) of the customer process it wants to support. In short, the values and cost are determined by various organizations performing roles that contribute to the value that is provided to the customer in the form of the e-service.

The following characteristics distinguish, according to Petrovic and Kittl (2003) a value net and determine its advantage compared to traditional businesses:

- *Customer-aligned.* Customer choices trigger the sourcing, building and delivery activities of a specific service or product within the value net. Distinct customer segments are offered customized solutions. Customers can play an active role. They are not passive recipients of supply chain output, but actively engaged in the service or production process in that they provide key information regarding their demands and personal preferences.

- *Collaborative, systemic*, and *information-based.* Companies engage suppliers, customers and even competitors in a unique network of value-creating relationships. Each activity is assigned to the partner best able to perform that specific activity. Significant portions of operational activities are delegated to specialist providers, for instance with regard to localization data, and the entire network operates smoothly thanks to collaborative, system-wide communication and information management. Information flow design and its intelligent use are at the heart of the value net. New digital information pathways connect and coordinate the activities of the value net, its customers and its providers. Rule-based, event-driven tools take over many operational decisions. Distilled real-time analysis enables rapid executive decision-making.

- *Agile* and *scalable.* Responsiveness to changes in demand, new product and service launches, rapid growth or re-design of the supplier network are all assured through a flexible production and distribution process, and information flows. Process time and steps are reduced. All processes in the value net, whether physical or virtual, are scalable.

Order-to-delivery cycles are fast and compressed. Rapid delivery goes hand in hand with reliable and convenient (electronic) delivery.

According to Selz (1999), the main characteristics of the organization domain model are cherry-picking from existing value systems, a value web broker acting as a central coordinator, an endeavor to come closer to the final consumer, and an integration of upstream activities, which is coordinated either with market platforms or with hierarchical mechanisms, or via a mixed form, i.e. networks (Powell, 1990). The value web model appropriates various concepts from transaction costs economics (Malone, 1987) and information systems theory. Markets, hierarchies, networks and information technology are woven into an intricate web of relationships to make value webs possible (Selz).

Within the value web, relationships between what we might call "structural" participants in the value networks are the most relevant. The balance of theory suggests that there are many reasons why firms would assume such structural roles – ranging from simple opportunism to requirements for new technological and market knowledge – but that the solidity of the relationship will depend largely on social and institutional antecedents. Depending on which actor(s) contribute key assets to the creation of value and the operating risks involved (Kothandaraman & Wilson, 2001), different configurations of actors are likely to result, some taking structural, integrative roles in the alliance and others taking supporting, facilitating roles. In deciding how to describe such a network is important to decide on the focal point of the value web and, starting from there, what the network looks like. It is clear that the description of the network or value web depends on the perception of the researcher, and it is difficult to delineate a network or value web and to decide which actor belongs either to the core or to the periphery of the network.

Although in reality, the lines between some of them may blur, we can identify at least three basic types of participants in any new value network (Hawkins, 2002):

- *Structural or tier-1 partners.* They provide essential and non-substitutable tangible and/or intangible assets to the value web on an equity or non-equity basis. They play a direct and core role in determining the intended customer value and in creating the business model.
- *Contributing or tier-2 partners.* They provide goods and/or services to meet requirements that are specific to the value web, but otherwise play no direct role in determining the intended customer value and in

creating the business model. If the assets they provide are substituted, the intended value and the business model could remain intact.

- *Support or tier-3 partners*. They provide generic goods and services to the value web, without which the value web would not be viable, but which otherwise could be used in connection with a wide variety of intended customer value and business models.

Structural partners make up the core of the network, while contributing and support partners connected to the network are loosely. As firms create products and services and engage customers in value exchanges, partners play an important role and require careful management (Galbraeth, 2002).

A remaining consideration in this scheme is the nature and longevity of the relationships involved. In principle, the assets and roles of contributing and supporting partners could be obtained in the wider market, through long-term or short-term contracts, depending on the circumstances. Many partners may only be required at specific points in time. Most structural partners would be in it for the long haul. Almost by definition, for the business model to survive, a structural partner leaving the alliance would have to be replaced by another partner bringing the same type of assets to the enterprise and playing the same role. A variation may occur when a structural partner's role is highly temporary – i.e. required to create and float the business model, but not essential to its subsequent operation. In such cases, it is likely that the assets contributed by this partner would be retained through a formal permission or license.

To conclude from a design perspective, specifically with regard to (mobile) internet services, access to critical resources is the key element in deciding which actors to incorporate in a value web. Critical resources for value webs are: access to the Internet and/or mobile infrastructure, to content, to content developers, aggregators and hosting providers, to software and application platforms, to customers, customer data, billing, customer support and management, based on the type of service providers of specific technology-related services, for instance mobile, location or positioning applications. Some of the resources may be found within a single organization, whereas for others more than one organization may be needed. Some resources may only be provided by one organization (structural partners), whereas for other resources several alternatives (support partners) may be available. A specific place is reserved for financial resources. Investments and the financial performance are crucially important in delivering value to customers and to the network.

Organization Design

The organization design describes the value network that is needed to realize the particular service offering. A value network consists of actors with certain resources and capabilities, which interact and together perform value activities, to create value for customers and to realize their own strategies and goals (see Fig. 2.5). Relevant topics in organizational design are:

- *Actors* can have greater or lesser degrees of power within in the value network, depending on the resources and capabilities they bring to the table. As we have seen, Hawkins (2002) identifies three basic types of partners in a value network: structural partners, contributing partners and supporting partners. In principle, structural partners are in a better position to exert control over the network than supporting partners are. Actors can fulfill single or multiple *Roles*, such as investor roles, governance body or technology provider.
- *Value Network.* The number of actors and the frequency and type of interactions contributes to the complexity and density of the value network.
- *Interactions and Relations.* Relations may evolve from reciprocal interactions. Relationships are important to a value network, because they contribute to trust and commitment within the network. Multiplexity refers to the number of levels within a relationship: the greater the number of levels, the stronger the relationship.
- *Strategies and Goals.* Actors vary with respect to the strategy and goals they pursue with the collaboration. Collaboration requires partners to share information and provide insight into each other's ways of working. However, strategic interest may induce partners to act against what is agreed upon, hide the truth or try to extract confidential information from their collaboration partners. Organizations may defend themselves by drawing up legal contracts and strictly monitoring another partner's activities. However, these safeguards do not guarantee that partners will not act opportunistically, which means that trust between partners is an important condition for an open and constructive collaboration.
- *Organizational Arrangements.* Collaboration leads to complex interdependencies between organizations, because no single partner has formal authority over another partner. Every adjustment has to be discussed and jointly agreed (Klein-Woolthuis, 1999). To govern the collaboration, actors need to agree formally and informally on how to

divide and co-ordinate their activities. These agreements should clearly define the responsibilities for all actors involved.

- *Value Activities* are the activities that an actor is supposed to perform in order for the value network to deliver the proposed service. A combination of value activities, together with the agreements and responsibilities, define the role an actor plays in a value network. Value activities can be seen as costs but also as a source of investment. If an actor performs a value activity and is paid directly for it, the activity can be seen as a cost. If the actor donates the activity in exchange for a part of the revenues later on, it is seen as an investment. We would argue that an actor who does not invest is not a structural actor.

- *Resources and Capabilities* can be financial, social, organizational and technical in nature. The technical resources and capabilities are the components with which the technical architecture is built. At the same time, the existing technical resources of the actors in the value network impose requirements on the technical architecture, as it has to be built using the resources that are available.

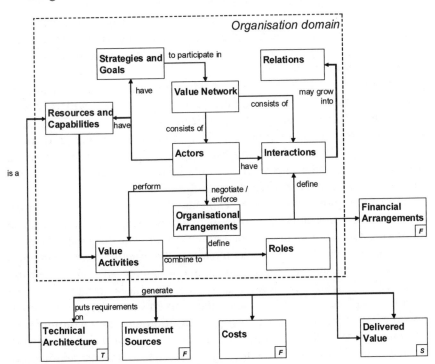

Fig. 2.5. Descriptive model for the organization domain

2.2.4 Finance Domain

Financial resources are one of the most important resources to be required by the value network. Finance also defines the bottom line of most of the services to be designed. With regard to financial arrangements, there are two main issues: investment decisions and revenue models. When it comes to investment decisions, there are a large number of surveys available (Demkes, 1999; Renkema, 1996; Van Oirsouw, Spaanderman, & De Vries, 1993). The authors of these surveys describe a number of methods that are predominantly based on financial criteria. They discuss general financial methods as well as multi-criteria, ratio and portfolio approaches (Renkema). Financial methods aim at average cost-effectiveness, net cash worth and internal return. Multi-criteria methods are those found in Information Economic, Kobler Unit Framework and the Siesta-method, which is partly based on the Strategic Alignment model. The ratio-methods are those found in Return-on-management and IT-assessment. Portfolio-methods are found in Bedell, investment portfolio and investment mapping (see Demkes; Renkema). Some methods go beyond the merely financial considerations, for example the balanced score cards (Kaplan & Norton, 1992, 1996), Value Prism (Neely, Adams, & Kennerley, 2002), while the option theory is a more detailed elaboration of the net cash worth concept (Demkes; Renkema). Demkes points out that decision-makers hardly ever use these kinds of methods.

Costs. Generally speaking, the cost side is reasonably well charted. The relative importance and absolute magnitude of cost drivers will vary from industry to industry, and from firm to firm. Exploiting and shaping these structural factors in defining the financial arrangements is very important. Drivers are partly related to internal relationships in a firm, partly to external factors, and partly to the relationship between internal and external factors (Stabell & Fjeldstad, 1998).

Transaction cost economics explains the firm's cost structure. Transaction costs include the costs of planning, adapting, executing and monitoring task completion. A transaction occurs when a good or service is transferred across a separable interface. Transaction cost economics identifies transaction efficiency as a major source of value, because enhanced efficiency reduces costs. The emphasis transaction cost economics places on efficiency may divert attention from other fundamental sources of value, such as innovation and the reconfiguration of resources (Amit & Zott, 2001). The business logic of the value network and the individual cost drivers form a framework to analyze and achieve an insight into the cost structure. Each firm faces the question whether to perform allocated tasks "in-house" or to outsource them.

The cost structure of most service businesses, including mobile services, is characterized by a high ratio of fixed to variable costs (Shapiro & Varian, 1999) and by a high degree of cost sharing such that the same facilities, equipment, and personnel are used to provide multiple services (Guiltinan, 1987). The high fixed costs typically lead to economies of scale, as increased production lowers the average production costs. Similarly, the high degree of cost-sharing leads to economies of scope, as the provisioning of a number of different services together leads to reductions in cost. Modularity in the service provisioning architecture is a way of obtaining cost advantage, as components or modules may be shared by several services. Service bundling, i.e. the combined offering of separate services as a package for a single price to customers, is another way to achieve cost advantage (Guiltinan). Complementarity between bundle components may stimulate demand, whereas cost sharing helps reduce costs.

Revenues. As far as the revenue side is concerned, which from our point of view includes realizing cost reductions as well as long term advantages that stem from intangibles, literature is less uniform (Low & Cohen Kalafut, 2002). Revenue models indicate what methods of payment are used, what is being paid for, and thus in what way income is generated. The thinking about models for income generation is less articulated than that with regard to business models. Furthermore, the distinction between the two is often vague. When talking about revenue models for the internet, Mahadevan (2000) distinguishes, for example, subscriptions, shopping mall operations, advertisements, computer services, general services, time usage and sponsoring (or free services). Weill and Vitale (2001) distinguish between (1) payments for transactions, (2) payments for information and advice, (3) payments for services and commissions, and (4) advertisement-generated income and payments for referrals. Holland, Bouwman, and Smidts (2001) discuss the following revenue models for Internet services: advertisement-based, transaction-based, models based on the float, subscription-based, licensed-based and models based on utility, i.e. "pay for" models. Rappa (2005), in addition to the revenue models mentioned above, considers infomediaries (selling customer data) and communities (sale of ancillary products).

Risks. Risks need some closer attention. Managers perceive risk in a way that is less precise and different from the way it is perceived in decision theory. The key point is that dealing with risk is a balancing act that should address both positive and negative aspects, taking the likelihood and subsequent consequences of any defined event into account. Telecom operators and service providers indicate that they are faced with a considerable number of risks and uncertainties concerning the technological and financial choices they have to make. Actors involved in the value web constantly examine the prospects of new technologies, technological possibilities and performance, and must decide whether or not to replace existing technologies. They are aware of the need to invest in new technologies and services, and although they may not know in advance when they may have to invest in new technologies, they have to be prepared to be able to catch up with their competitors quickly enough when the time arrives. Other risks associated with implementing new technologies are the risks regarding the availability of the new technology, standards (will the technology be standardized to allow for mass production to occur), the risk of irreversibility of the chosen technology path, the risk of the chosen technology becoming outdated and path dependency (future technological choices depend on former paths set in). To determine the practical impact and risks involved, all these issues need to be taken into account.

Pricing. Revenues depend on the price associated with the service. In its simplest form, the price of a service is the amount of money a customer has to pay for using that service. In an extended definition, price refers to all the sacrifices the customer has to make to obtain and use the service. In telecommunication services, switching costs can be considerable. Pricing, i.e. setting prices for a product or service, is a dynamic process that takes internal and external factors into account, e.g. cost considerations and competition from alternative services.

Different pricing strategies may be used for new innovative products or services. Providers may skim the high-end market for short term profit, or try to create market share aiming at long term profit (Kotler, 1999). Price setting then includes pricing of core products, and optional and necessary complementary products. Price for data transport and content can be charged individually. Charging for data transport may be session-based, volume-based, usage-based flat rate or package-based. Content pricing can be pay-per download, subscription-based event-based, context-based or value-based. Charging for content services is becoming more and more advanced, and there is a shift towards new models like context-based and value-based pricing. This is quite different from the traditional approaches used in the "mobile voice business". The former is a form of cost-based pricing, i.e. setting a price such that costs are covered with an acceptable margin. This approach has a fundamental problem in that allocating fixed costs depends on sales volumes that are intrinsically linked to prices. The second approach is competitive pricing, i.e. prices are set based on similar offerings in the marketplace. The danger here is that it may stimulate the notion that services are commodities, and expose the entire industry to price wars (AT Kearney, 2003).

As far as consumers are concerned, there is no relationship between the production costs of a service and its value. Unlike service providers, consumers are simply not interested in the costs of generating the service, but only in the value the service represents to them. This means that the price of a service may be based on the consumer's perception of its value. This is called "perceived-value pricing" (Kotler, 1988) or value-based pricing (Jonasson & Holma, 2002). Since consumers may vary in their perceptions, value-based pricing is quite challenging (Klein & Loebbecke, 2003).

Price discrimination, which refers to charging different prices to different customers for the same goods or services (Shapiro & Varian, 1999), is also widely applied. Shapiro and Varian distinguish three forms of price discrimination (1) personalized pricing, (2) group pricing, and (3) versioning. Klein and Loebbecke (2003) add (4) bundling, and (5) price discrimination based on sales volume. Empirical data shows that differential pricing is already widespread in industries that are characterized by high levels of fixed costs, like the airline, telecom or publishing industries (Varian, 1996). In the telecom industry, for example, differential pricing may be based on customer characteristics, e.g. preference for pre-paid or post-paid, group characteristics, e.g. student discounts, product characteristics, e.g. tailored bundling of specific services and features (e.g. device, voice and data services), and volume (the more minutes you call the cheaper the price per minute). In the airline industry, business travelers are typically time-sensitive and leisure travelers typically price-sensitive. To differentiate between business travelers and leisure travelers effectively, the airline industry offers cheaper tickets only if a weekend is included in the trip (Klein & Loebbecke).

Pricing plays a different role in throughout the life cycle of a service. It may be impossible to charge for a service or required equipment in the initial phase, e.g. when the value of the service increases with the size of the installed base. This initially leads to "give-away strategies", such as those observed in e.g. mobile telecommunications.

An important question is how investments and revenues are arranged and shared within complex value networks. Important stakeholders in complex value systems are close to the core or structural actors, actors that invest, i.e. banks, or investors that make the roll-out of new products or services by small starting companies possible, i.e. venture capitalists. Investment decisions reflect the interests of the actors involved and take the mutual benefits of multiple organizations into account. Organizations that are connected through intended relationships and interdependencies consider sharing risks, solving common problems, acquiring access to complementary knowledge to be major motivators for collective investments. To facilitate interorganizational investments, organizations go through a collective decision-making process. Compared to internal processes, these joint processes have the following implications (Demkes, 1999). They require a lengthy decision-making process, demand multiple rounds of negotiations, while dealing with conflicting interests (not always resulting in a win-win situation for all parties concerned). Investment in which multiple actors are involved, incur high transaction costs and generate disputes.

Interorganizational investments require explicit articulation and collective agreement on the terms of investment and timing (Miller & Lessard, 2000). The share of each participant and the corresponding partnership ratio must be defined. It will be determined what each member will contribute in terms of financial and technical expertise. The success of these arrangements hinges on whether or not the role of each member within the terms of institutional framework is clearly defined (ibid.)

Finance Design

The finance design describes the financial arrangements between the various actors in the value network. It shows how the value network intends to capture monetary value. For a business model to be viable, the division and sharing of benefits and costs should be balanced to create a win-win for the involved partners. The structure of the value network has a strong influence on the financial variables resulting in archetypical financial arrangements (see Fig. 2.6).

- *Investment sources.* The investments and costs are closely related to the design choices made in the technology design. However, the question as to who will supply *Capital* is another important design variable in the finance domain.
- *Cost sources.* The costs may also be influenced by the coordination costs of the value network.
- *Performance indicators.* With regard to the evaluation and management of the financial arrangements over time, performance indicators, like market adoption, usage, return on investment, et cetera, are necessary.
- *Revenue sources.* Revenues can come directly from the end-user, but there may also be other sources of revenue (for example subscription, advertisement or government subsidy).
- *Risk sources.* The risks that may exist in the other domains have financial consequences. For example, if the perceived customer value is much lower than the assumed value, this may have a negative impact on

the revenues. The way the value network copes with the financial consequences of the various risks is part of the financial arrangements. Scenarios for investments, costs and revenues may be used to counter future risk and uncertainty.

- *Pricing.* The price and the pricing structure is the most visible part of the arrangements as far as the end-user is concerned. Possible solutions are dependent on pre-pay or subscription based pricing, flat free charge, or per usage pricing.
- *Financial arrangements.* The financial arrangements between the actors in the value network describe the way profits, investments, costs, risks and revenues are shared among the actors. These agreements should clearly define the benefits for all actors involved.

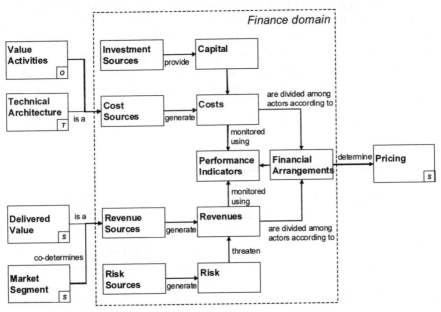

Fig. 2.6. Descriptive model for the finance domain

Based on the analysis of the four domains and the specific issues discussed for the four domains, it is possible to analyze and design business models. However, we believe that business models, are not a one time result, but change all the time, and are dynamic in nature.

2.3 From a Static Model to a Dynamic Model

Business models are dynamic in nature, with design choices having to be adapted over time in order to maintain a fit with the environment. External factors, such as socio-economic trends, technological developments, and political and legal changes, often require business models to be adapted over time. In addition, endogenous business model dynamics occur as changes in one component of the business model may require adaptations in other components to maintain an internal fit. And finally, the different stages in which services are developed require design choices to be adapted in order to be aligned with the characteristics of the phase in the business model life cycle.

In order to gain insight into how business models change over time, it is helpful to structure the life cycle of business models in phases (cf. Afuah & Tucci, 2001). We will begin by discussing such a phased model for business models, and then take a look at the forces in the environment of companies that force them to make alterations to their business models. Finally, we discuss how we use business model components, phases and external forces as the building blocks for discussing the dynamics of business models.

2.3.1 Business Model Phases

The development process from business idea to established business can be divided into a number of phases (Kubr, Marchesi, Ilar, & Kienhuis, 1998). Phasing models helps to understand the evolution of the competitive landscape in the wake of an innovation or change, and the consequences of such events for firm strategies and business models (Afuah & Tucci, 2001). The phases to which we refer here are conceptualized in several disciplines. We have briefly examined phasing from five perspectives: innovation management, technical service development, entrepreneurial and business planning, innovation adoption and diffusion, and marketing. When we compare the phases as distinguished in these perspectives, broadly speaking there are three main phases: the technology/R&D phase, the implementation/roll-out phase, and the market phase, where the market combines the sub-phases of market offering, maturity and decline. To get a better view of these phasing concepts, we provide a brief description below from various perspectives.

Innovation management perspective. According to innovation management literature, new products and services follow a life cycle from innovation to development and maturity. Next, they enter a phase where new generations of products and services overtake the by now mature innovation. Rapid and frequent product and service innovation is prevalent in the early phases, while later stages are characterized by stable product and service concepts, with only incremental changes motivated by cost reduction and cost-effective exploitation (Tidd, Bessant, & Pavitt, 2001). A similar phasing model is proposed by Veryzer (1998), who observes that the wild exploration of new technologies leads to a converged product formulation and further specifications. After the specified product is evaluated, it goes through increasingly specific prototyping stages, after which the product is commercialized.

Technical service development perspective. According to Räisänen et al. (2005), the technical service development lifecycle includes functionalities needed to provide service from development to retirement. They distinguish the following activities as part of the service life cycle: service creation, service provisioning and service lifecycle management in general during (during the complete lifecycle). The service provisioning level can be divided into service deployment, service usage, service retirement and service operational management. At the lowest level of abstraction, Räisänen et al. distinguish the following activities at the service usage level: service discovery, service selection, service negotiation, service composition, service adaptation, service execution and service termination.

In each of these phases, there are important hurdles to overcome. In the service creation phase, technical factors are the most important, whereas in the service provisioning phase, extra attention should be paid to environmental factors like legislation and market adoption. It is important to emphasize that, in practice, the process is not sequential but iterative. It can be seen as a loop in which deployed services may be optimized on the basis of on feedback from an operational network as well as other input.

Entrepreneurial and business planning perspective. Also, from entre-preneurial and business planning perspective, the development process from business idea to established business can be divided into different phases. Burgelman (1983) describes four processes that may overlap and be iterative: defining the activities in the innovation project; gaining support within the organization; the structural implementation of the new product; and embedding the innovation in the company strategy. Another popular model is the one suggested by Mason and Rohner (2002), who distinguish four venturing phases:

- Phase I – venture vision (validating the concept): the objectives of this phase are putting a plan together that outlines the product or service and its uniqueness, the market and why it is attractive, the team and why they are qualified, and a high-level business model and the amount of money needed.

- Phase II – alpha offering (building while planning): in this phase, the objectives are to build the alpha version of the product/service and its platform, see how it works, factor in changes and refinements to the basic specifications, and understand and resolve implications for the venture's positioning, its business model, value propositions, as well as the effects on functional strategies and plans of the rest of the venture's departments.
- Phase III – beta offering (testing the concept): in this phase, the objectives are to test and refine the product or service and (by implication) the rest of the venture's program, gain early market acceptance and customer testimonials from a beta product, and to use beta stage results to secure funding to do a full market launch. Testing the business model thoroughly is an important key skill in this phase.
- Phase IV – market offering (calibrating and expanding): the objectives of this phase are to find customers, become profitable, and get the next version of the offering to market.

The model provided by Mason and Rohner (2002) is primarily aimed at how an organization can develop and offer a new service. The Internet technology life cycle model as described by Afuah and Tucci (2001) adopts a broader view by phasing the process in which companies in an emerging industry in general (like the Internet industry) are developing their services and business models. They distinguish the following three phases:

- The emerging or fluid phase: in this phase, (many) new entrants as well as incumbent players choose their profit sites and value network positions. There is competition between new and old technologies and between different designs using new technology. Product quality is low, costs and prices are high, market penetration is low, with mostly lead users and high-income users as customers. Since product/service and market requirements are still ambiguous in nature, there are few failures in this phase.
- The growth or transitional phase: in this phase, a standard or dominant design defines a critical point in the life cycle of the innovation. The customer base moves to a mass market. Competition and disappearing ambiguity with regard to requirements force many firms to exit or make important changes to their business models. The firms with the best adapted, and the most viable and feasible business models survive.

- The mature or stable phase: in this phase, companies focus on keeping and improving their competitive advantages. In markets where imitation is easy (e.g. most (mobile) Internet services), companies continuously apply (incremental) innovations to their business models.

Innovation adoption and diffusion perspective. Rogers (1995) discusses the adoption and diffusion of an innovation in a social system and defines it as a process with three phases: adoption, diffusion and maturity. In the first phase of the process, only a small portion of the members of a social system (the innovators) will adopt the innovation (and the rate of adoption is still low), but once the early adopters have joined in, the innovation adoption curve rises steeply (the diffusion process is gaining momentum), and by the time the late majority has also adopted the innovation only the laggards are left, who will take considerable time before embracing the innovation. Then, after the adoption and diffusion of the innovation, the maturity phase has been reached.

Marketing perspective. The most popular phasing concept from marketing point of view is the product life cycle (PLC) concept, developed by Theodor Levitt in 1965 (Kotler, 2000; Moon, 2005). Based on the assumptions that products have a limited life time, product sales pass through distinct stages (each posing different challenges, opportunities, and problems to the seller), profits rise and fall at different stages of the product life cycle, and that products require different marketing, financial, manufacturing, purchasing, and human resource strategies in each stage of their life cycle, he distinguished four stages: introduction, growth, maturity, and decline. The life cycle stages as described by the more strategically oriented Johnson and Scholes (2002) are quite similar to that of Levitt. The stages they distinguish are development, growth, shake-out, maturity and decline. The shake-out stage can be seen as a "slowing growth phase" between growth and maturity, in which users and buyers are increasingly selective with respect to buying services and the weakest competitors disappear from the market.

There are critics who argue that in reality phases are not fixed. Instead, there is variability in shape and duration (Kotler, 2000). In addition, life cycles need not be sequential in practice. Adopting a non-sequential, iterative view instead of a linear view of life cycles may prove to be valuable: by versioning or repositioning products or services, their actual life time can be extended (Moon, 2005). Making these types of changes may also lead to changes in the underlying business models.

Business model phasing. Based on existing phasing models, we propose an analysis that includes three phases in the life cycle of business models, namely (1) Technology/R&D, (2) Implementation/Roll-out, and

(3) Market. The start of the Technology/R&D phase is triggered by the initial conceptualizations of the service concept (the solution to a problem) and business model. In this phase, R&D (basic and applied research) and technology play a dominant role, the core activities being service or product definitions, investment in new technologies and collaboration with technology providers. The transition from the first to the second phase is marked by the launch of the service on the market (start of the Market phase). Activities in this phase are testing service concepts in focus groups, field experiments, the roll-out of technology, testing of alpha and beta versions of the service, and (small-scale) roll-out on the market. The business model moves from the second to third phase once the service reaches critical mass after market experiments prove successful. Core activities in the market phase are retaining rather than capturing market share, commercial exploitation on a day-to-day basis, focusing on operations and maintenance. As such, the market phase subsumes stages of market offering, maturity and decline. Although our description of these phases suggests a linear process, feedback loops may occur, especially in the early phases, when business models do not develop as planned.

2.3.2 External Forces

Maintaining a "fit" with external factors is important in keeping a business model sustainable over time (Morris et al., 2005). Hughes et al. (2007) demonstrate that external constraints on business models can be technical, economic, cognitive, structural, legal, political and cultural in nature. According to Hill and Jones (1995), there are two types of environments that can influence the performance of firms: the industry or competitive environment and the macro-environment. At the industry level, Porter's (1985) model of competitive and industry analysis is relevant. At the macro-environmental level, factors identified in the PESTEL-framework are relevant, i.e. political, economic, social, technological, environmental, and legal factors (Johnson & Scholes, 2002). In our STOF model, we summarize the competitive and macro-environments into market drivers, i.e. influence of suppliers, customers and competitors, technology drivers, i.e. influence of changing technology and innovations, and regulation drivers, i.e. privacy, intellectual property rights, competitive and other kinds of regulation, that have the most direct impact on the business model.

2.3.3 A Dynamic Business Model Framework

The results from the literature review of business models, value networks, phasing and dynamics come together in our dynamic STOF model as depicted in Fig. 2.7, which links the four STOF business model components, the external forces, and the three phases we identified.

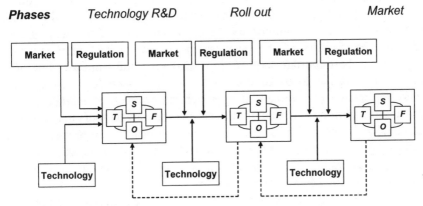

Fig. 2.7. An overview of phases from several disciplines

Now that we have a clear idea of the three core phases of a business model development path, we can take a look at the causalities between phases, external forces and business model components. The impact of external drivers on internal business model components will be different in each phase. In the Technology/R&D phase, technology will be the major driver behind new business model development. Specifically, the emergence of new mobile, wireless and data networks, like the Internet, help to increase the reach of businesses, while at the same time middleware, web services and multimedia applications offer new opportunities for enriched, customized and secure communication. However, new business models can also be driven by market developments, such as changing customer demand or new entrants on the market, or by regulatory changes, such as liberalization or changes in, for instance, interconnection regimes that allow for kick back fees. With regard to the service concept, technical and organizational arrangements, as well as finance a large number of issues as discussed before have to be dealt with in this first phase.

In the Implementation/Roll-out phase, regulators and competitors become aware of the new product and services, and they will look into possible regulatory implications and prepare a strategic response, for instance by demanding stricter regulations. As a result, it has to be certain that the service complies with regulation regarding such issues as fair competition, privacy, intellectual property rights, and content restrictions. Changes in market factors, for instance competitors copying the service concept, and technology, i.e. the availability of more innovative or cheaper solutions, can affect the service and business model, but we expect them to be less conditional in nature than regulation, and have less of an impact on the business model in this phase. Regarding the internal components, experimentations in the Roll-out phase lead to information on operations of technology and perceived ease of use and utility as experienced by lead users that will have an impact on the customer value of the service and therefore the viability of the business model. We expect branding and scalability to be important, both from a marketing and from a technology perspective. As the focus of the activities shifts from R&D to a more market-oriented and commercial approach, assets such as market know-how and promotion become more important, and new partners may need to be included in the organizational network, for instance to reach specific market segments. As far as the financial component is concerned, practical issues such as pricing and bundling of service offerings need to be solved.

Finally, in the Market phase, we expect market drivers to play an important role, as the focus is on retaining customers and competitors are starting to revise their business models in response to the service offering. Regulation is aimed more at the surveillance of existing rules to ensure they are observed, and technology drivers have a minor impact, as the internal technology business model component mainly deals with scalability, operations, and maintenance, requiring periodical updates rather than full technology architecture revision. With regard to the service component, questions concerning the delivered value, customer satisfaction and customer retention become relevant. Organizational component issues deal with process optimization and operational management. With regard to financial issues, relevant issues are commercial revenue generation, maintenance and operation costs, oriented towards effectiveness and cost reduction.

The discussion of the dynamic STOF model, and the way it is embedded in literature regarding services, technology, organization, strategy, innovation studies, forms the basis of our design approach, the Critical Design Issues and the Critical Success Factors that we will discuss in the next chapter. In Chap. 8 we will specifically discuss business model dynamics.

Chapter 3 STOF Model: Critical Design Issues and Critical Success Factors

H. Bouwman, E. Faber, E. Fielt, T. Haaker, and M. De Reuver

In this chapter we will make the transition towards the design of business models and the related critical issues. We develop a model that helps us understand the causalities that play a role in understanding the viability and feasibility of the business models, i.e. long-term profitability and market adoption. We argue that designing viable business models requires balancing the requirements and interests of the actors involved, within and between the various business model domains. Requirements in the service domain guide the design choices in the technology domain, which in turn affect network formation and the financial arrangements. It is important to understand the Critical Design Issues (CDIs) involved in business models and their interdependencies. In this chapter, we present the Critical Design Issues involved in designing mobile service business models, and demonstrate how they are linked to the Critical Success Factors (CSFs) with regard to business model viability. This results in a causal model for understanding business model viability, as well as providing grounding for the business model design approach outlined in Chap. 5.

3.1 Designing the Business Models

Based on the theoretical and the core technological issues discussed so far, we derived the first part of our STOF model, i.e. descriptive models for the service, technology, and organization and finance domains respectively, in Chap. 2. These models contain the most important design variables in each domain. We use the term 'design variable' to denote that our model focuses on variables that can be influenced by design teams, business developers and managers.

A central element in our model is the fact that a viable business model should create value for customer and network alike. Creating customer value is not an easy task due to the difficulty of extracting user requirements, conflicting design requirements and a lack of the resources needed to provide perfect solutions. Although design choices in the technology domain should in principle satisfy the design requirements of the service domain, not every solution will be affordable, which means that there are also interdependencies between the technical domain and the financial domain.

Creating value for business actors (network value) is complex due to the conflicting strategic interests of partner organizations. Actors often originate from different industries (e.g. network operators, financial institutions and retailers), and each have their own strategic interests (e.g. generate traffic, extend services to customers, generate transactions). Design choices in the organization and finance domains may serve the strategic interests of the actors involved.

In the business model literature, knowledge on how effectively balance the requirements and strategic interests within and between the different domains is largely missing. To develop insight into how organizations can design 'balanced' business models, designers need to understand the critical design issues (CDIs) in business models and their interdependencies. A CDI is defined as a design variable that is perceived to be (by practitioner and/or researcher) of eminent importance to the viability and sustainability of the business model under study.

To elaborate our approach towards the design of viable business models, we present the CDIs for mobile services' business models in the next section.

3.1.1 Critical Design Issues

We have identified the common and recurrent CDIs from a large number of case studies involving the business models of mobile services. The descriptive STOF model we discussed in Chap. 2 has been used to describe and analyze the business models of a number of cases, including mobile entertainment services, mobile tracking and tracing services, (mobile) community services, presence and instant messaging services, business to employee services and mobile payment services (see Table 3.1).

Table 3.1. Overview of cases

Services type	Cases	Reference
Mobile entertainment services	My Babes, Radio 538 ring tones	Maitland, Van De Kar, When De Montalvo, and Bouwman (2005)
Mobile tracking and tracing services	TMC4U (Traffic Management Channel for You), Traphic SMS alerts (Vialis), Finder i-mode service (KPN Mobile)	Faber, Haaker, Bouwman, and Rietkerk (2003) and Maitland et al. (2005)
(Mobile) community services	I-Karos, Vaccination database, Botfighter	Rietkerk and Timmerman (2003)
(Mobile) Presence and instant messaging services	Splendo, Jaytown, MSN	Kijl and Timmerman (2003)
Business to employee services	P-info, Lucio, Zorgpas, Caremore	Bouwman and Van Den Ham (2003b)
Mobile payment services	Moxmo, Mobile2Pay, Mobipay	Faber and Bouwman (2003)

The aim of the case studies was to identify CDIs. Based on the case study descriptions (see Table 3.1 for more extensive publications on individual cases), specific CDIs were extracted for every domain, and then clustered systematically. Based on the recurrence of issues and/or their perceived relevance with regard to the viability of the business model, as indicated by the interviewees and coded as such, these issues were qualified as critical.

We will discuss the CDIs in more detail, establishing what the CDIs in each domain are, and how they relate to balanced business models. Next, knowledge on CDIs is used to build causal models describing the interrelatedness of design issues and their relationship with business model viability (see also Haaker, Faber, & Bouwman, 2006).

Critical Design Issues in the Service Domain

CDIs that originate from the service domain based on the cases analyzed are *Targeting, Creating Value Elements, Branding* and *Customer Retention.*

- *Targeting.* An important issue in almost every case was choosing a profitable target group. Should the service offering be targeted at consumers or businesses? Should the service offering be targeted at a niche market or at a mass market? Should the service focus on youngsters or elderly people? et cetera. Sometimes, service providers formulated a growth strategy in which the target group evolved from one market segment to another. For instance, the strategy used by Moxmo, a payment service provider, was to extend its activities and the related target group from micro-payments (e.g. ticketing and ring tones) to medium-sized payments (e.g. payment of compact discs).

- *Creating Value Elements.* Closely connected to choosing a target group is formulating a compelling value proposition for end-users. The added value of a service can be based on value elements like fun, efficiency, accuracy, speed, personalization, trust, et cetera. It is the augmented service offering, i.e. the core service plus the support or auxiliary services, which creates the customer value. For example, the availability of a helpdesk was found to have a significant influence on customer value in the case of Caremore, a service that supports homecare professionals with administrative and informational services. The cases indicate that there is a clear tension between the possibilities offered by technology and the wishes and needs expressed by end-users. Quite a number of the studied services did not have a clear and compelling value proposition and seemed to be blinded by the technical possibilities. For instance, P-info (see also Chap. 9), a mobile service for police officers to access information in police databases, suffered from a poor fit with prevailing working practices, a cumbersome interface and limited access to relevant databases. Instead of using the new mobile service to retrieve information, police officers preferred to use the existing voice-based service via the control room. In several cases, an important value element was trust. The main question was how to enhance trust (or reduce the need for trust). In the cases we studied, the objects of trust varied. In some cases, trust had to do with the reliability (branding) of business actors (e.g. mobile payment cases), whereas in other cases trust was associated with the security and privacy of the technology that was deployed (e.g. presence and instant messaging cases). Also, different mechanisms were used to enhance trust. For instance, in the mobile payment cases, we found that Mobipay used a trusted third party (institution-based trust), whereas Moxmo relied primarily on recurrent positive experiences (process-based trust).
- *Branding.* Branding was found to be an important issue in relation to reaching the customers that had been targeted. Brands seem to have a direct influence on the perceived value of service offerings, which makes it an important means to create customer value. Brands were used for different purposes in the cases we studied. First and foremost, brands were used to increase the visibility of services in the market (all cases). Secondly, brands were also used to communicate value elements, such as trustworthiness (e.g. mobile payment and community cases) or image (Radio 538 ring tones). Providers may decide to promote a new service by

bundling it with an existing product, which carries either the same brand name or a different brand name. The TMC4U traffic information service, for example, was promoted and sponsored by a manufacturer of Travel Management Channel (TMC) modules, although it did not carry the sponsor's name. In the case of Radio 538 ring tones, the popular Radio 538 brand was explicitly linked to the service to increase service awareness and communicate service image. An important requirement for brand choice is brand recognition by the target group and the existence of a match with the intended value proposition.

- *Customer Retention.* In addition to choosing a target group and defining the added value, customer retention was found to be a CDI. Customer retention refers to marketing strategies aimed to keep customers satisfied and loyal with the product or service. The cases show that service providers adopt different strategies to stimulate recurrent usage of their services. In the entertainment cases, versioning, service bundling and personalization were used to promote customer retention. Personalization, accuracy and actuality of information were used in the tracking and tracing cases to attract and retain users. In the cases involving presence and instant messaging services, the strategy was to introduce new versions with new functionality. Finally, in the business to employee service Caremore, customization of the service was used to enhance the service's value for the employee to stimulate loyalty to the service.

The extracted CDIs and related design requirements are summarized in Table 3.2.

Table 3.2. Critical Design Issues and related design requirements (service domain)

Critical design issue	Description	Balance of requirements
Targeting	How to define the target group?	Generic vs. niche service B2C vs. B2B service
Creating value elements	How to create value for the targeted users of the service?	Technological possibilities vs. user needs and wishes
Branding	How to promote/brand the service?	Operator vs. content brand
Customer retention	How to stimulate recurrent usage of service?	Customer lock-in vs. customer annoyance

Critical Design Issues in the Technology Domain

Based on our case studies, CDIs that originate from the technology domain are *Security, Quality of Service, System Integration, Accessibility for Customers*, and *Management of User Profiles*.

- *Security.* Trust of end-users and customers in a service offering is partly determined by the way security is implemented in the technical architecture. That is, the way in which access to a service is granted and how security of communication and (stored) information is realized. Security often requires a trade-off between ease of use or privacy considerations and preventing abuse. For example, in the community and instant messaging cases access ranges from anonymous access, use of a username (nickname) and password (MSN Messenger), to full user identification (enterprise Presence and Instant Messaging – PIM – service). With anonymous access, privacy is guaranteed, but users cannot be traced in case of abuse of the service. In the case of mobile entertainment services, authentication of users is simply based on the SIM card in their mobile phone. Security may be realized more easily in a closed environment. For example, in the case of enterprise PIM services, the company may deploy its own Instant Messaging (IM) server shielded from the outside world by a firewall. Obviously, the service cannot include contacts from outside the company, thereby limiting its use and value.

- *Quality of Service.* In all the cases we studied, the performance of the technical architecture in delivering the technical functionalities has a profound impact on the service offering and perceived value. A balance between the quality of the service and the incurred costs has to be maintained. A typical performance measure that influenced the quality of service in the tracking and tracing cases was the accuracy of the deployed positioning technology. Likewise the value of provided information depends on the actuality of the information. As far as mobile entertainment and business to employee services are concerned, the speed of the service, for example the download time of content, was found to be an important parameter. Availability of the service under changing circumstances was important for business to employee services like P-info and Caremore.

- *System Integration.* The adoption of the service is in part determined by the extent to which the new service can be integrated into the existing technical infrastructure. The trade-off with system integration is between flexibility and costs. The costs for building on legacy systems may be lower but provides for less flexibility than an open system based on standards and open interfaces. For instance, for mobile payment services the degree of integration with existing payment solutions is an important barrier for merchant adoption. In the Botfighter case the Geographic Information System (GIS) was not integrated in the general platform but included in the specific Botfighter application, as no generally accepted standard for the GIS was available at that time. In principle, the use of open standards and architectures allows for easier integration between systems.

- *Accessibility for Customers.* The accessibility of the service to the target group is obviously influenced by the choice of platforms, devices and architecture. For example, a closed architecture reserves service access to a restricted target group. This may be intentional, for instance in the case of business to employee services and enterprise instant messaging services, or in the case of mobile entertainment services offered exclusively by an operator to its customers. However, a service may be unavailable unintentionally when a service requires specific resources (compliant handset) or capabilities (coping with cumbersome user interface) from the end-user. For example, the TMC4U service requires a specific TMC module. Users can only access the service if they invest in such a module coupled to their car radio. Adoption of the Caremore service required substantial training of personnel in order to be able to cope with the mobile device. Similarly, the adoption of the P-Info service (a mobile data service for police officers) was hindered because officers had a strong preference for voice interfaces and access to critical databases was not realized.

- *Management of User Profiles.* With regard to the personalization of a service, a user profile that contains user interests, preferences and behavior must be created and maintained. The management of this profile, i.e. the creation, use, maintenance and access to the profile, requires technical functionality that may be realized in different ways. There has to be a balance between user involvement and automatic profile generation, and between privacy and access to the user's profile. For MSN Messenger the instant messenger server keeps a profile for each user. A privacy statement is issued to users regarding the protection of the data they provide. In the case of the Traphic SMS-alert service, the user controls his profile, containing times and routes of travel, via Internet. In the i-mode Finder case, the necessary location information is determined automatically by the operator and transferred anonymously to the location-based service provider, after the user has given his consent.

The technology domain's CDIs and design requirements are summarized in Table 3.3. Note that the CDIs in the technology domain are often enablers for the CDIs in the service domain.

Table 3.3. Critical Design Issues and related design requirements (technology domain)

Critical design issue	Description	Balance of requirements
Security	How to arrange secure access and communication?	Ease of use vs. abuse and privacy
Quality of service	How to provide for the desired level of quality?	Quality vs. costs
System integration	How to integrate new services with existing systems?	Flexibility vs. costs
Accessibility for customers	How to realize technical accessibility to the service for the target group?	Open vs. closed system
Management of user profiles	How to manage and maintain user profiles?	User involvement vs. automatic generation

Critical Design Issues in the Organization Domain

Based on our case analyses, CDIs that originate from the organization domain are *Partner Selection, Network Openness, Network Governance,* and *Network Complexity.*

- *Partner Selection.* An important design issue in all cases is acquiring access to resources and capabilities needed to realize a service offering. Firms need to decide whether to outsource certain activities or to perform them in-house. A distinction can be drawn between business actors that provide indispensable and irreplaceable (critical) resources and capabilities, and those who provide supporting resources and capabilities. For instance, in the traffic information cases (Traphic SMS alerts and TMC4U), an important issue was whether or not to include the government in the value network. Given the cost of acquiring and processing raw traffic data, government funding is considered a critical resource for any 'commercial' traffic service. In the mobile payment cases (Moxmo, Mobipay and Mobile2pay), an important issue was whether or not to include a financial institution as transaction enabler and trusted third party, in the value network. Whereas Moxmo decided to operate independently from the financial institutions to reduce transaction costs, Mobipay and Mobile2pay decided to include one or more financial institutions in the value network to enhance trust (see also Chap. 10). Access to critical resources and capabilities (e.g. customers, content, funds, et cetera) was found to be an important strategic interest when selecting partners.

- *Network Openness.* The level of openness indicates the degree to which new business actors can join the value network and are allowed to provide services to customers. In our cases, we observed two different organizational arrangements: the closed model, in which a relatively fixed consortium of partners collaborate, and the walled garden model, in which new partners are able to join the value network if they comply to certain rules. In the entertainment cases (My Babes and Radio 538 ring tones), the i-mode Finder case, and community cases (Ikaros, Vaccination database, and Botfighter), for example, portal providers used a walled garden model to control the quality of the provided content. In the presence and instant messaging cases we found instances of a closed model (Splendo News messenger and Jaytown). No instances were found of an open model in which partners are free to join the value network and offer services and content. When choosing between various degrees of network openness the desired control, exclusiveness and customer reach of the service were found to be of main strategic concern. The higher the desired level of control and exclusiveness is, the more likely partners are to adopt a closed model. On the other hand, reaching many customers may be an argument in favor of choosing an open model.
- *Network Governance.* In all of the cases, we found a dominant actor, often the one with access to the customers and end-users or the one that developed the service offering, which was managing the value network. These business actors often approached and selected collaboration partners, set the rules with regard to collaboration (organizational arrangements), and monitored compliance with these rules. For instance, in the entertainment cases, community cases, and the PIM cases that focused on B2C applications, the portal provider is the dominant actor, while in the business to employee cases and some PIM cases, e.g. MSN messenger, it was the application service provider who acted as the dominant actor. Typically, actors with access to customers shield these relationships from other actors in the value network. Customer ownership thus seems to be of key strategic concern to actors in the value network.
- *Network Complexity.* The cases we studied vary with respect to network complexity. Network complexity may arise from the number of relationships a focal business actor needs to manage in a value network and from the effort needed to connect the actors' IT applications and systems (technical architecture). We found that business actors tend to reduce network complexity by using intermediaries, which act as single points of access. In the i-mode Finder case, for example, we found that

the portal provider (network operator) chose to reduce network complexity by using an intermediary actor to manage the relations with the different content providers. In the Zorgpas case we found that the high number of organizations that needed to collaborate (20) resulted in an enormous network governance load and efficiency losses. Finally, Mobipay's transaction platform for mobile payment requires the acceptance and collaboration of all major financial institutions in Spain. Hence, Mobipay is faced with a considerable degree of network complexity. Moxmo, on the other hand, chose to bypass the financial institutions for its service offering, in order to reduce network complexity. There is a trade-off between the need to reduce complexity and the need to have access to critical resources and capabilities.

The extracted CDIs and related strategic interests are summarized in Table 3.4.

Table 3.4. Critical Design Issues and related strategic interests (organization domain)

Critical design issue	Description	Strategic interests
Partner selection	How are partners selected?	Access to critical resources and capabilities
Network openness	Who is allowed to join the value network?	Desired exclusiveness, control, and customer reach of service
Network governance	How is the value network orchestrated? Who is the dominant actor?	Customer ownership and control over capabilities and resources
Network complexity	How to manage increasing number of relations with actors in a value network?	Controllability of value network and access to resources and capabilities

Critical Design Issues in the Finance Domain

CDIs that originate from the finance domain are *Pricing, Division of Investments, Division of Costs and Revenues*, and *Valuation of Contributions and Benefits*.

- *Pricing.* With regard to the adoption and actual use of a service, the perceived customer value must at least equal, and preferably exceed, the price of a service. In Mobile Payment case examples, the service is free of charge and even entitles user to reduced prices for purchased goods. The aim is to attract and retain customers. The traffic information service TMC4U offers traffic messages via the RDS channel on car radios. It is free of charge. However, to appreciate this service as a truly personalized service, the driver needs to invest in a car navigation system equipped with a TMC module. The service is sponsored by a provider of navigation systems and TMC modules. The user of Traffic SMS alerts pays a premium SMS price. The service is characterized by relatively high variable costs and virtually no fixed costs for end-users. The i-mode services (Mybabes, Radio 538 ringtones, Finder) use identical pricing mechanisms: they require users to invest in an i-mode phone, operator subscription, i-mode subscription, flat fee service subscription and fees depending on data traffic. The height of the fees, including those for the services offered by third parties, is set by the dominant actor, in this case the operator. Pricing seems to be above all aligned with the aims of the (dominant actor in the) value network, e.g. it is aimed at maximizing profits or creating market share.

- *Division of Investments and Risks.* There are financial risks involved in developing and introducing a new service, as there is uncertainty about the resulting return on the investment. In the case of Caremore, a business to employee service for homecare professionals, some of the uncertainty was reduced by adopting a phased approach. Prior to the actual roll-out of the service, it was tested in pilot groups. Traffic information services like TMC4U and Traphic SMS alerts rely on the government to make large investments in infrastructure for acquiring and processing raw traffic data. In the mobile entertainment cases (My babes, Radio 538 ringtones), the content providers are responsible for the investments needed to provide content in a format that is acceptable for the operator. Nordic operator Telia introduced a location-based game (Botfighter), which was targeted at the youth segment. Telia regarded the investments in the game as a means to win the (long term) loyalty of the youth. However, to reduce the upfront investment Telia did not develop the game itself. This was done by It's Alive!, which in return receives a monthly fee plus a share from the SMS revenues. The division of investments seems to match partners' profitability and risk profile.

- *Valuation of Contributions and Benefits.* For fair and viable revenue sharing arrangements it is important to value the contribution of each partner to the service offering and the (intangible) benefits each partner receives. In the Caremore case, for example, the choice in favor of a specific operator was based on an existing trust relationship, and on the quality of network coverage. In the same case the appreciation of the system integrator changed over time. When the health organization providing Caremore acquired the necessary competences for system integration itself, the initial external systems integrator was considered too expensive and no longer included in the value network. For Microsoft the benefits from its (free) MSN Messenger are mostly intangible: it ties users to the portal and software of Microsoft. Some revenues are obtained from its link with SMS services. However, less than 5% of the revenues of Messenger-SMS's is distributed to Microsoft. Most of the revenues (90%) go to the network operator whose payment relation with the customer provides a strong position in the value network, apparently resulting in this large percentage. The main goal of the provider of SMS traffic alerts to start the service was to learn about the market for traffic information and getting access to customers on this market. For this benefit the provider was even prepared to incur a small financial loss. It appears that the valuation of contributions is based on the access actors have to resources. The intangible benefits to the actors depend on their strategic interests like e.g. acquiring market knowledge or access to customers.

- *Division of Costs and Revenues.* We found that the division of costs and revenues was different from case to case and that in each case it may follow a different logic, e.g. cost based or value based. In the i-mode case the operator and the content partners split the revenues from the fixed monthly service subscription fees. There is no clear relationship between the costs and revenues of the content provider. Other cases show a connection between incurred costs and revenues. The operator in the Mobile entertainment cases, for example, received revenues based on data transport. In other cases there is a relationship between invested money and share of revenue, for example in the Botfighter case game developer It's Alive and platform provider Ericsson receive a percentage of the SMS revenues. In some cases (SMS Traphic alerts, MSN messenger), although the revenues the dominant actor receives are lower than the costs, the dominant company feels sufficiently compensated through intangible benefits. The relationship between costs and revenues for each of the actors involved seems to depend on their individual access to critical resources, the valuation of these resources, the risks and level of investments, and the existence of intangible benefits.

The extracted CDIs and related strategic interests are summarized in Table 3.5.

Table 3.5. Critical Design Issues and related strategic interests (finance domain)

Critical design issue	Description	Strategic interests
Pricing	How to price the service for end-users and customers?	Realize network profitability Realize market share
Division of investments	How to divide the investments among business partners?	Match individual partners' profitability and risk
Valuation of contributions and benefits	How to measure and quantify partners' contributions and (intangible) benefits?	Fair division of costs and revenues
Division of costs and revenues	How to divide the cost and revenues among business partners?	Balance between individual partners' profitability and network profitability

3.2 Critical Success Factors for Customer and Network Value

The viability of a business model is determined by the creation of value for the customers and for the organizations in the business network. The CDIs are instrumental in the value creating process as they serve as the starting-point for our causal model. First, we will identify CSFs with regard to creating customer and network value. Wohltorf and Albayrak (2005) describe a decision support system for aiding service design, which is based on success factors for technical feasibility of the service and economic viability. CSFs refer to "the limited number of areas in which satisfactory results will ensure that the business model creates value for the customer and for the business network" (adapted from Rockart & Bullen, 1981). Secondly, we will formulate design propositions that describe the relationships between the CDIs and CSFs, creating a break-down structure that explains how business model viability can be influenced by more concrete, design-oriented variables that influence the success factors regarding creating customer and network value.

3.2.1 Creating Customer Value

A number of CSFs with regard to viable mobile service business models exist as requirements for creating customer value. A *Compelling Value Proposition* (service domain) is an obvious requirement when it comes to creating customer value (Edvardsson, Gustafsson, & Enquist, 2006; Weill & Vitale, 2001). A value proposition refers to the benefits that are delivered to the user of a service by its provider. This requires focusing on what creates value from the point of view of the customer (Edvardsson et al.) instead of the possibilities of the technology. The CDI *Creating Value Elements* (service domain) enables a compelling value proposition. Moreover, a compelling value proposition is determined by CDIs like *Branding* (service domain) and *Pricing* (finance domain). Customers perceive a brand as an important element of a value proposition, and it can be used to differentiate the value proportion from those of competitors (Kotler, 2000).

A second CSF is a *Clearly Defined Target Group* (service domain). This enables the service provider to stay focused on the customers (Edvardsson et al., 2006). Market segmentation is a compromise between the unrealistic assumption that all customers are the same and the uneconomic assumption that all customers can be treated as individuals (Kotler, 2000). This also means that having a *Compelling Value Proposition* and a *Clearly Defined Target Group* are interrelated. The CDI *Clearly Defined Target Group* (service domain) deals with choices between consumer or business market and between niche or mass market. The result is a clearly defined target group. The possibilities regarding targeting are co-determined by the *Accessibility for Customers* (technology domain). Here we see that a CDI originating from the technology domain can influence a CDI originating from the service domain.

Unobtrusive Customer Retention (service domain) is also a CSF for customer value. Although firms often strive for customer retention, this can reduce customer value when obtrusive mechanisms are used. Obtrusive mechanisms can hamper the ease-of-use (Davis, 1989) and create negative experiences that frustrate users (Strauss, Schmidt, & Schoeler, 2005). Therefore, the CDI *Customer Retention* (service domain) should lead to unobtrusive mechanisms, the design of which can be supported by *User Profile Management* (technology domain) that makes it possible to personalize of the service.

While the previous three CSFs relate to the service domain, the fourth CSF is related to the technology domain. An acceptable *Quality of Service* delivery (technology domain) is required because, as far as services are concerned, the quality of the service process (functional quality) is as important as that of the service outcome (technical quality) (Grönroos, 1994). Because mobile services are delivered via technology, the CDIs that are related to the technology domain should lead to an acceptable quality level. The *Quality of Service* relates to the performance of the technological architecture in delivering the functionality. The *Security* deals with the access to the service and the security of communication and information processing. The compatibility refers to the level of *System Integration* with the existing technical infrastructure, and that between subsystems.

To conclude, the CSFs for creating customer value are *Clearly Defined Target Group, Compelling Value Proposition, Unobtrusive Customer Retention,* and an *Acceptable Quality of Service.* It is assumed that high scores on these success factors will result in a service that meets the user expectations, i.e. a service that generates customer value (Fig. 3.1). A service that generates customer value in the long run can be expected to result in a viable business model.

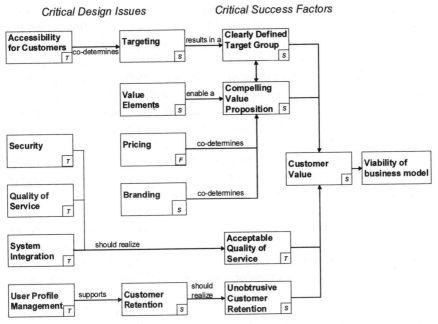

Fig. 3.1. Critical design issues and critical success factors relating to creating customer value

3.2.2 Creating Network Value

There are also requirements when it comes to creating network value. In the business network firms will, on the one hand, cooperate to create value based on common interests and, on the other hand, competing to capture value based upon individual interests (Brandenburger & Nalebuff, 1996). Whereas some authors emphasize competition, for example Porter's five forces model (Porter, 1980), others emphasize cooperation, such as industrial marketing and purchasing (e.g. Axelsson & Easton, 1992). CSFs with regard to network value relate to balancing these forces in the finance and organization domains resulting in acceptable outcomes for the participating firms, in particular those firms that provide essential resources and capabilities.

Because financial incentives are important to ensure the participation of firms in new business initiatives, the profitability and risks for the firms in the business network are CSFs. Experiences with respect to electronic business have taught us that paying not enough attention to 'the bottom line' results in the failure of new business initiatives (Holland, Bouwman, & Smidts, 2001). An *Acceptable Profitability* (finance domain) should be acceptable in an absolute sense, that is to say a positive financial result matching companies' risk/return profile, as well as in a relative sense, that is compared to the financial results of the other participating firms. The CDI *Division of Costs and Revenues* (finance domain) should result in an acceptable level of profitability. *Acceptable Profitability* is determined by the *Pricing* and an *Acceptable Customer Base* (service domain). An acceptable customer base depends upon CDIs related to *Customer Retention* (service domain), *Accessibility for Customers* (technology domain), and *Network Openness* (organization domain).

Acceptable Risks (finance domain) is a CSF with regard to mobile initiatives because of the high uncertainty with respect to market acceptance and technology choices (Haaker, Faber, & Bouwman, 2006). The CDI *Division of Investments* (finance domain) should result in acceptable financial risks. The *Division of Investments* and the *Division of Costs and Revenues* are both enabled by the CDI *Valuation of Contributions and Benefits* (finance domain). The measurement of tangible and intangible benefits is essential to realizing value from IT investments (Ward & Daniel, 2005).

However, financial factors are not the only kind of CSFs required to generate network value. Organizational factors also need to be taken into consideration. A *Sustainable Network Strategy* (organization domain) is required for securing access to (inimitable) resources and capabilities, including capabilities for managing the network (Gulati, Nohria, & Zaheer, 2000). The latter is referred to by Ritter, Wilkinson, and Johnston (2002) as network competence: "a company-specific ability to initiate, handle, and use inter-organizational relationships." The CDI *Network Governance* (organization domain) influences the sustainability of a network strategy.

An *Acceptable Division of Roles* (organization domain) refers to the distribution of roles among firms and the integration of roles within firms that participate in the business network. Kambil and Short (1994) have already drawn attention to the importance of roles and their connections to the functioning of business networks. This is also related to the decision to insource or outsource. Outsourcing involves the decision to place activities outside a firm in order to economize on resources (service level versus costs) and/or acquire capabilities, allowing a firm to focus on its core competencies (Smith, Mitra, & Narasimhan, 1998). The CDI *Partner Selection* (organization domain) determines the way the roles are divided. The CDI *Network Complexity* (organization domain) influences both the sustainable network strategy and an acceptable division of roles.

In conclusion, the CSFs with regard to network value are *Acceptable Profitability, Acceptable Risks, Sustainable Network Strategy*, and an *Acceptable Division of Roles*. It is assumed that high scores on these success factors will result in a 'win–win' situation, in which all actors have an incentive to participate, i.e. a business model that generates network value (Fig. 3.2). It can be expected that a service that generates network value in the long run will result in a viable business model.

The CDIs in the organization domain are instrumental for dividing value activities among multiple actors and aligning their resources, capabilities and strategic interests. Similarly, the CDIs in the finance domain are instrumental in defining financial arrangements that lead to a profitable business for all parties involved.

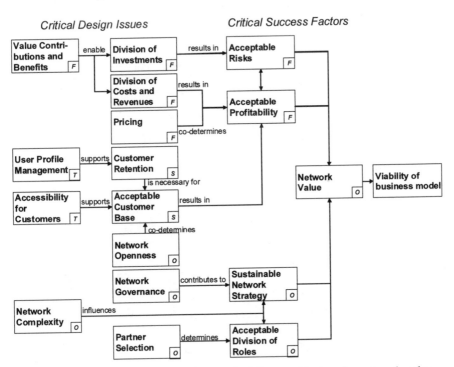

Fig. 3.2. Critical Design Issues and Critical Success Factors for network value

Together, Figs. 3.1 and 3.2 form a causal model that explains the viability of business models based on the assumption that a viable business model should provide both customer and network value. In this causal model, the CDIs are the instruments used by designers and managers to influence business model feasibility and viability.

In Chap. 5, a process model for designing business models will be presented.

Chapter 4 The Mobile Context Explored

M. De Reuver, H. Bouwman, and T. De Koning

There is a difference between designing mobile services and designing other electronic services. The specific characteristics of the mobile services domain shape the context in which design choices on business models are made. Whereas the previous chapters discussed service innovation and business model development in general, in this chapter we take a closer look at our core concepts in relation to the mobile services domain.

We begin with a brief discussion of the main characteristics that distinguish mobile services and business models from other electronic services. Next, we describe the trends that are relevant to each of our four business model components, i.e. mobile services offered and adopted; technology trends; resources, roles and models in the organization domain; and financial models, as used within the mobile sector. Finally, we discuss typical classifications of mobile services business models and compare them to other e-business model classifications.

4.1 The Value of Mobility

While the use of other electronic services is bound to a fixed location, mobile services allow users to consume the service anytime, anyplace. This mobility is prominent in the definition of mobile services we provided in Chap. 1, i.e. they assume mobility from the user of the services, the devices, the sessions or applications, and they may be offered via mobile and wireless networks. It is important to note that mobile services include, but are not limited to, services that are offered over wireless networks. For example, a music file can be side-loaded through the Universal Serial Bus (USB) connection of a fixed desktop computer, and then consumed while on the move at a later moment. Similarly, a mobile payment service may use Radio Frequency Identification (RFID) technology, which works at a short range but does not provide mobility management. The mobility aspect enables mobile extensions of existing electronic services, as well as new services that are specifically valuable to users on the move. On the other hand, the potential obtrusiveness of mobile services is higher, and there are technical issues with regard to maintaining the quality of the service experience in different environments that may have to be resolved.

Another differentiator is the personal nature of mobile devices as a result of the fact that most people have their own, personal mobile device. There is almost no other item that is carried everywhere by users in quite the same way. This symbiotic relationship makes it possible to identify users and collect data about their demographics, handset type or typical behavior, which can be used to personalize service experiences and thus strengthen this symbiotic relationship. However, this also means that issues regarding privacy, security and identity management become even more important for mobile services than they are for other electronic services.

In addition to static personal information, real-time user-related context information can make mobile services more useful and relevant. Context information can be any information that is relevant to the use of a service. One example is location information, which is typically generated through Global Positioning System (GPS) or network triangulation. Other types of context information are information about the time of day, air temperature, device battery power, tasks in the user's agenda, social contacts, or even blood pressure and heart rate. Context-related information allows for automatic service delivery at the relevant time, the tagging of user generated content, and more efficient service usage. However, the use of context-related information also raises design issues with regard to privacy and security.

Although the abovementioned characteristics of mobile services seem to be fairly straightforward, Arnold (2003) discusses the mobile phone as an innovation with paradoxical consequences, explaining that, while the mobile phone is wireless and portable, it is fixed as users require a fixed number to be reachable. And while the personal nature of the mobile phone makes it a private device, its frequent use in the public sphere makes it public as well. Moreover, while the mobile phone brings users closer to the persons they communicate with, it also creates a distance between themselves and the people around them.

Compared to fixed networks and desktop computers, wireless networks and devices pose various technological challenges to service developers. With regard to the networks, data rates are often lower than fixed networks, while the costs per data packet are higher. Handheld devices generally possess less processing power, less available memory and limited battery power, which puts a limit on the use of high-end applications and compression technologies. Due to the small screens and keyboards, web-content and navigation techniques have to be adapted, which is a complex affair because there are many different types of handsets, operating systems and micro-browsers.

While in the case of fixed Internet, there is almost no formal regulation, it plays an important role in the mobile services domain. Often, licenses are required to operate wireless networks, because the available spectrum is limited. Other current topics in regulation include the unbundling of value chains and network neutrality. At the level of middleware, applications and services, however, there is hardly any mobile-specific regulation.

From an organizational point of view, an important difference is the high level of dependency between the actors involved in offering services. While Internet Service Providers (ISPs) merely provide connectivity, cellular network operators also control access to the customer and billing services, and impose rules that content providers should adhere to when offering services over their networks. In this chapter, we take a closer look at the roles and relationships in mobile services value networks.

The characteristics discussed in this section are some but not all of the issues that make mobile services different from other electronic services. In the next sections we discuss the service, technology, organization and finance-related specificities and trends of the mobile services domain in greater detail.

4.2 Mobile Services

Although there is a large variety of mobile services being offered in the marketplace, as far as the consumer market is concerned, most of the more advanced services have not yet been widely adopted by end-users. SMS messaging, search services, ring tones and icons continue to be the most popular services (Bouwman, Carlsson, Molina-Castillo, & Walden, 2007). While many users have a phone to access the mobile Internet, only a fraction of them actually uses theirs for that purpose (Forrester, 2006a), with the exceptions of Japan and South Korea, where there are more mobile Internet users than wired users. Longitudinal research conducted in Finland (Bouwman, Carlsson, et al., 2007) shows a gradual increase in the use of mobile data-services. The most important services in this respect are mobile e-mail and Internet surfing, the use of which gradually increases over the years. Travel and mobile commerce-related services are less well-established. Few users are as yet familiar with context and location-aware services. Ironically enough, some of the services that initially attracted much attention as possible killer applications, like Mobile TV, stock trading and mobile adult content, are now among the least popular services. Analysis of these services makes it clear that services really have to fit the behavior of consumers and business people. The adoption of advanced mobile services is most likely limited to specific niche market segments, and generally speaking is associated with gradual growth rates.

In the remainder of this section, we discuss several types of services that are enabled by mobile technologies.

Information services. Typical information services include search services, news and weather, transportation timetables, and yellow pages. Information can be easily modified, consumed time and again by the same or different users, and reproduction is fast and cheap. Typically, information on the mobile Internet includes (Barnes, 2002):

- Text, e.g. news, stock prices, film listings, advertisements, product descriptions and restaurant locations
- Audio, e.g. voice, wireless Internet radio and music files (including MP3 format)
- Graphics, e.g. wireless bitmap or Graphics Interchange Formats (GIFs)
- Video, e.g. animated graphics files, mobile and wireless TV and video files

Mobile information services can be delivered through text messaging (i.e. SMS), Multimedia Messaging Service (i.e. MMS) or mobile Internet. Mobile Internet making use of Wireless Application Protocol (WAP) was launched in 1999 by the Japanese operator DoCoMo. Their so-called i-mode service provided a portal to the content of several providers. Billing and connectivity are both provided by the operator, while the unified look and feel of the content gives end-users a coherent user experience. i-mode has been copied by similar concepts like Vodafone Live!, T-Zones and Orange World. However, the adoption of these services outside of Japan has been disappointing. One might argue that the success of i-mode in Japan is to a large extent due to contextual factors like low Internet adoption, low PC penetration and long commute trips.

In addition to these on-portal content models, several off-portal WAP sites have emerged that provide information services. The most successful among them are those that offer content downloading services for mobile device personalization, such as ring tones and wall papers.

Location data can be used to make information services more useful, e.g. by filtering the content that is being distributed to the user. In addition, they also enable new types of services, including navigation services that provide driving directions or maps, and tracking and tracing services to find the location of family members, friends or objects (Van De Kar, 2004).

Communication and messaging services. Mobile data services can complement voice communication services. The most popular of these services has been person-to-person SMS, i.e. short text messaging. While initial expectations regarding this service were low, it has proven to be the most successful service to date. More advanced messaging using pictures and video data is made possible through MMS services. Currently, video telephony services are being heavily promoted by 3G operators. The use of mobile e-mail is also increasing, mainly via specialized devices like the Blackberry. A potential disruptive innovation as far as SMS is concerned is instant messaging or group messaging. Instant messaging is a solution that has been adapted from the fixed Internet that uses presence information. In contrast to SMS, users do not pay for every message they send, but only for the data traffic they generate.

Entertainment services. Entertainment services include downloading music, watching television, playing games, jokes, horoscopes, gambling, and chatting. A special type of mobile entertainment services is mobile TV. While entertainment is often developed by professional artists or news agencies, there is a growing trend in user-generated content, driven by camera-enabled phones for making pictures and videos. As a result, users become active contributors and content developers rather than mere passive consumers.

Transaction services. Another type of services is related to transactions and payments. Mobile payments involve 'wireless transactions of a monetary value from one party to another using a mobile device [...] over a wireless network' (Ondrus & Pigneur, 2005, p. 1). These could be micro-payments, e.g., for transportation tickets or vending machines, or macro-payments, e.g., for movie tickets, shopping or restaurants (Mallat, Rossi, & Tuunainen, 2004). A related type of services is mobile banking, i.e. accessing information on account balances and carrying out transactions (Mallat et al., 2004). Mobile remittance services are particularly relevant in developing countries because they make it possible to transfer money between individuals in the absence of financial institutions (see Chap. 13).

Business services. In addition to consumers, businesses can also benefit from using mobile services. Examples of business services include mobile sales-force automation, mobile supply-chain management (SCM), mobile access to email, Personal Information Management (PIM) applications, mobile tracing and tracking, and mobile dispatching and scheduling (Wang, 2007). Benefits associated with the adoption of these applications include reductions in travel time and greater flexibility in working environments. In addition, back office processes can be streamlined through automatic and on-the-spot administrative processes. Businesses can also adopt services aimed at localizing and exchanging data with goods, for example by using RFID solutions. A specific type of business service are emergency support services. Police officers, for example, can benefit from mobile devices that provide access to back office systems, or they can be alerted when there is an incident in their vicinity that warrants their attention. Also, fire and ambulance staff may carry devices with geographical information with regard to risks and proper procedures.

Each of these services provides a specific type of added value, e.g., entertainment, information, transactional, or communication. One of the challenges facing companies in the mobile domain is to find the specific niche market segments at which to aim their services. Designing a service that adds value and matches the behavioral pattern of consumers or businesses is crucially important with regard to successful mobile service innovation. Even when a company manages to design a successful service, however, chances are that it will be adopted gradually rather than swiftly.

4.3 Mobile and Wireless Technology

In this section, we explain the major trends and characteristics of technologies that enable mobile services. To begin with, we focus on wireless networks, describing recent developments as well as future trends. Next, we discuss middleware and applications that enable mobile services and mobile Internet access. We conclude with a brief discussion regarding handsets.

4.3.1 Wireless Networks

Over the past 15 years, the wireless industry has witnessed a dramatic technological evolution. Starting from analog, voice-only, circuit-switched transmission, today's networks provide high-speed, packet-switched, digital voice and data services (Kumar, 2001). There are two types of wireless access technologies. Cellular networks cover large areas up to the size of countries. Other wireless access technologies, on the other hand, often provide higher data rates but cover smaller areas. These short-range technologies are typically used to cover smaller, densely populated areas, such as city centers or university campuses. Larger coverage can be gained by configuring them to support hand-offs between antennas or by integrating them with cellular networks.

Wireless access networks are often classified on the basis of two dimensions: the data rate for end-users and the range of the network, see Fig. 4.1. Typically, there are trade-offs between these dimensions: while cellular network technologies offer high ranges they offer modest data rates, whereas other technologies, such as WiFi, offer higher data rates but at the cost of shorter range.

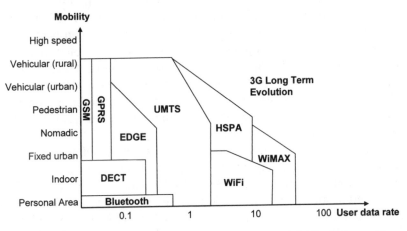

Fig. 4.1. A comparison of wireless access technologies

Deploying a cellular network requires high investments, including the costs involved in deploying antennas and the underlying infrastructure, as well as the costs involved in acquiring a license for using spectrum. Because cellular networks tend to reuse existing antenna sites and core network assets from previous generations, cellular networks are typically controlled by large, international network operators, making it difficult for new players to enter the market.

Short-range access networks, on the other hand, are easier to deploy. The costs involved in setting up antennas and transceiver stations are typically lower, and as they cover smaller areas, initial investments in the network involved are lower. In addition, no licenses are needed for many of these technologies, which means that there are no costs or hurdles involved in deploying a network.

Cellular Networks

When the first generation of cellular networks was deployed in the 1980s, it provided analog voice telephony, and adoption rates were low. Cellular networks became popular when the second generation networks were rolled out. In European countries, the GSM (Global System Mobile) standard was an important driver, while cdmaOne, which was used in the USA, apparently, had less of an impact. These were the first technologies providing digital voice telephony to rival the quality of wired networks. In addition, GSM introduced international roaming, i.e. being able to use the telephone in foreign countries.

Although GSM provides high quality voice telephony, it is less usable for data services, because data rates are too low. In addition, the circuit-switched transmission is inefficient, because it occupies a fixed amount of bandwidth during connection. As a result, users are charged based on the time they are connected to the network rather than the amount of data that is actually transmitted. To overcome these problems, GPRS (General Packet Radio System) and EDGE (Enhanced Data rates for GSM Evolution) were developed. In addition to providing the higher data rates needed for Internet browsing, these technologies offer packet-switched transmission, allowing users to be 'always connected'.

The third generation mobile networks was expected to deliver true broadband quality. Currently, there are two competing standards: Wideband Code division multiple access (W-CDMA), such as Universal Mobile Telecommunications System – UMTS – in Europe, is the follow-up of GSM, while CDMA2000 builds on cdmaOne networks. In addition to offering slightly improved data rates, the main advantage of UMTS is that it allows prioritization of traffic according to desired quality of service.

Because the capacity and data rates of UMTS are insufficient to allow efficient broadband access, evolutionary improvements have been developed. High-Speed Downlink Packet Access (HSDPA) provides higher download data rates, an increase in the capacity of the antennas, and a reduction in the response time of the network. This is complemented by High-Speed Uplink Packet Access (HSUPA), which increases upload capacity. When the two are combined, data rates can be achieved that are comparable to wired Asymmetric Digital Subscriber Line (ADSL) networks. Standardization of follow-up evolutions are currently taking place, which will probably simplify network architectures, increase data rates, and reduce latencies and costs per data packet (*UMTS Forum*, 2006).

An important development with regard to cellular networks is IP Multimedia Subsystem (IMS), which is part of the UMTS standard. IMS is a central network component that can integrate cellular networks with fixed networks and short-range access networks such as WiFi. Because integration takes place within the cellular operator's network, the operator remains in control of the end-user (Cuevas, Moreno, Vidales, & Einsiedler,

2006) and allows operators to limit access to public Internet pages (Braet & Ballon, 2007).

Short-Range Access Technologies

In parallel to the evolution involving cellular networks, short-range access technologies are emerging, mostly originating from the fixed Internet domain. The best known of these technologies is Wireless LAN or WiFi, a wireless extension of the Ethernet standard that can typically reach around 50 m. There are multiple WiFi standards, providing 11 Mbit s^{-1} up to over 100 Mbit s^{-1}. The next generation of WiFi, 802.11n, is expected to provide even higher data rates. While the technology does little more than offer a wireless access point to a fixed network, access points can be combined in order to cover larger areas. Compared to cellular networks, WiFi is faster, cheaper and more efficient when deployed in densely populated areas. However, security, interference and a lack of wide area coverage are drawbacks. In practice, WiFi is used for in-home Internet access, corporate voice over IP and Intranet, public access points to the Internet, and peer-to-peer networking (Bohlin, Lindmark, Rodríguez, & Burgelman, 2006).

The mobile version of WiMAX (Worldwide Interoperability for Microwave Access) is a new technology that offers ranges of up to a few kilometers. Compared to WiFi, it offers higher data rates, wider coverage and improved security. In addition, it allows for a more efficient use of the spectrum than WiFi does. On the other hand, it is more expensive to deploy and less suited for local high-speed coverage. In addition to providing Internet access, WiMAX can be used to cover the gaps left in WiFi network access points. It can also be combined with cellular networks to increase the capacity of these networks.

Several personal area network (PAN) technologies exist that are typically used for machine-to-machine communication (Frodigh, Parkvall, Roobol, Johansson, & Larsson, 2001). These short-range technologies include Bluetooth, Ultra Wide Band and Zygbee. Bluetooth was originally developed to replace cables at low costs and low power. It uses the same unlicensed spectrum as WiFi and can cover between 10 and 100 m at low data rates. Ultra Wide Band is not yet widely adopted because of potential interference. It uses very low power pulses in a high range of the spectrum and can reach about 20 m.

A special application of short-range technologies is the use in ad hoc networks, which are formed dynamically by wireless mobile nodes, without using a centralized administration or network infrastructure (Chlamtac, Conti, & Liu, 2003; Niemegeers & Heemstra De Groot, 2003). If enough users are present, reliability and user mobility could be high.

This technology eliminates the need for network components like base stations and central controllers. As a result, ad hoc networks are typically quicker and cheaper to deploy than centralized networks. In addition, they are more flexible and theoretically more reliable, as there is no dependence on a central network component. Disadvantages are the lack of standard routing protocols, security issues and a reduced reliability when there are fewer users.

4.3.2 Accessing the Mobile Internet

There are various middleware and applications that enable mobile devices to access the Internet. The first attempt to enable mobile Internet was WAP, the purpose of which was to enable the easy and fast delivery of relevant information and services to mobile users. Although WAP-based browsers never met these expectations, the transport mechanism is still in use (Jaokar & Fish, 2004). The central part of the WAP architecture is the WAP gateway between the application server and the mobile device. The gateway communicates with the application server using regular Hypertext Transfer Protocol (HTTP) traffic. However, with the end-user client it uses Wireless Mark-up Language (WML) to overcome the limitations associated with HTTP on the wireless Internet (Jaokar & Fish). WAP can also be used to push Internet content to end-users by using the WAP push protocol, i.e. by including the link in an SMS message. The WAP protocol has been the basis for i-mode, which is a proprietary technology designed for browsing WAP sites on a mobile device.

An alternative to accessing applications on a remote server is to run applications on the handset itself using Java 2 Micro edition (J2ME) or Binary Runtime Environment for Wireless (BREW). Extended Hypertext Markup Language (xHTML) mark-up is used for J2ME to adapt the content to fit the screen of the device.

Mobile Web Services

A promising new approach to realizing mobile Internet access is web services technology (Farley & Capp, 2005; Pashtan, 2005). In recent years, web services have become increasingly popular in the IT world. Web services are the underlying tools of the Service Oriented Architecture (SOA), which defines three basic roles. The service requestor is the party requesting to receive a service from an external party. The party providing the service is called the service provider. The services that a provider offers are published in the database of the service broker, and a service requestor can find the right provider for his needs by querying this broker. Communication between service provider and requestor takes place via asynchronous messages. Web services define a set of protocols that can be used to implement SOA. The SOAP-standard (Simple Object Access Protocol) is used to send messages between service requestors and providers. The messages are built up using the XML protocol (Extensible Markup Language). Web Service Description Language (WSDL) is the standard being used to describe the services published by the service provider. WSDL files are stored by the service broker using the UDDI protocol (Universal Description, Discovery and Integration). All these standards are open, meaning that any service provider and requestor can use web services to establish mutual interaction. The standards also 'shield' the internal complexity of the application hosted by the service provider, simplifying communication between providers and requestors, and guaranteeing interoperability between their systems. Because it is fairly easy for service requestors to switch to other services provided by other providers, web services are very flexible. Another advantage is that services offered by providers are reusable, i.e. a service offering's generic functionality that is usable for multiple end-user services can be used over and again in many service compositions.

As mobile communication networks increasingly offer high speed data communication, a logical next step would be at least to consider applying web services technology to mobile devices. This approach is known as mobile web services (MWS) (Farley & Capp, 2005). There are many benefits associated with MWS compared to the way service development and execution currently take place. Because MWS uses open standards, it is much easier to develop new services, and it is easier to reuse services (Farley & Capp, 2005), the latter of which is especially relevant to generic service components that can be integrated into other services, such as authentication, billing and context information services. In addition, MWS will enable dynamic service discovery and is expected to benefit the interoperability of mobile devices (Pilioura, Tsalgatidou, & Hadjiefthymiades, 2003). Given the many different mobile devices and the diversity of operating systems and parsers, interoperability is currently not guaranteed, which leads to cost inefficiencies. MWS is also expected to benefit end-users, as the ease of developing services will probably lead to a greater variety and number of services, a higher level of personalization, better and simpler user interfaces, and the provisioning of services over the most appropriate access technology (Farley & Capp, 2005).

To illustrate the potential of MWS, some applications can be considered. A problem that MWS could solve is that users may have multiple mobile devices such as phones, PDAs and laptops, and that the user's contacts are typically not stored in all of them. To solve this, the contacts could be stored in a central database, and users can then access the data via MWS (Farley & Capp, 2005). Another application of MWS is to use the technology for providing generic service components, such as geographical maps or authentication and billing services. Instead of having to develop these components for various service offerings, service providers can reuse and integrate them into their specific service offerings. Generic authentication and payment services are in fact the main services advocated in the MWS architecture proposed by Microsoft and Vodafone (*Microsoft*, 2003). In addition to requesting (generic) services from the outside world, MWS can also be used to enable external parties to request information from the mobile device. For example, a context-aware application offered by a service provider may request information about a user's location or current occupations. If standard web services published by the user in a UDDI are used to this end, any service provider could, in principle, easily add context awareness elements to his value proposition. In a similar approach, services could request context data from portable sensors, contact information, payment information or other personal information from the user (Berger, McFaddin, Narayanaswami, & Raghunath, 2003).

Although the benefits of MWS are clear in terms of flexibility, interoperability and reusability, there are also several challenges mentioned in literature when it comes to applying web services in a mobile context. One obvious difference between a mobile and a wireline device is the portability and the mobility of the device, i.e. a mobile device can move to another location. In addition to mobility, the personal nature of a mobile device also differentiates MWS from fixed Internet web services. A mobile device is generally owned by one person only, which makes it possible to link the device to a specific user identity. This offers opportunities, for example with regard to personalizing the MWS. However, it also raises issues involving identity management and privacy that are more serious in the mobile setting than they are in fixed web services (Farley & Capp, 2005).

We also identify a set of challenges related to the specific performance limitations of both mobile devices and infrastructures. Although we agree that some of these issues may become less urgent as technology advances, MWS today have to be designed in such a way as to limit their use of computation, memory and energy resources (Berger et al., 2003). Another issue involved here is that the user interface with small screen and keyboard makes it more difficult to fill in forms, which has an impact on the usability of MWS (Farley & Capp, 2005). Also, because bandwidth is still smaller and more expensive in mobile networks than it is in the fixed

Internet, limited data rates need to be taken into account. Issues regarding bandwidth and processing power are especially challenging for MWS, as web services use XML and SOAP protocols to code the messages. Processing XML messages on mobile devices requires higher levels of processing power compared to HTML messages (Limbu, Wah, & Yushi, 2004). In addition, XML messages generate a larger overhead than HTML pages, as they can reach up to five times the size of content messages (Tian, Voigt, Naumowicz, Ritter, & Schiller, 2004). In cases where bandwidth is indeed limited, or where users are charged per packet of transmitted data, this may well become a constraining issue.

Although standardized MWS architectures have not yet been defined, industry parties like Microsoft and Vodafone (*Microsoft*, 2003) are working together in developing a MWS architecture, as well as Nokia and Sun (*Nokia & Sun*, 2004). The open Internet standardization body OMA (Open Mobile Alliance) is also working on a standardized MWS architecture. An important trade-off in these efforts involves the choice between using open protocols to implement MWS or using proprietary standards. Of course, using proprietary standards may well put inter-operability at a risk, and may lead to walled garden alike business models.

4.3.3 Devices

While early mobile phones offered only voice calling and text messaging functionalities, the current trend is towards multifunctional consumer electronics devices. At the moment, high-end mobile devices offer radio/music/video players and editors, Internet browsers, agenda functions, voice recorders and cameras. In addition to offering GSM access, most devices nowadays include Bluetooth and Infrared functionalities, and they can increasingly access 3G and WiFi networks. While mobile phones are taking over these traditionally separated consumer electronics functions, there is also an opposite trend that involves MP3-players being equipped with cellular communication functionality, e.g., Apple's iPhone.

In parallel, processing power, data storage capacities, and screen resolution have increased tremendously. Having said that, there are considerable differences between high-end devices like Personal Digital Assistants (PDAs) and smartphones on the one hand, and regular mobile phones on he other. In terms of software, today's devices are capable of executing Java applications, browsing the Internet through specialized mobile phone browsing software, and have specific operating systems, e.g. Microsoft Windows Mobile and Symbian.

Enabling technologies for mobile and wireless services are developing rapidly. The capacity, data rates and cost effectiveness of wireless networks are increasing continuously. In addition, increasing functionality of software components and intelligence of devices are changing the landscape of involved actors in the mobile domain.

4.4 Organizational Arrangements

In the era of plain voice telephony, the mobile telecommunications value chain was fairly simple, consisting only of network operators and equipment manufacturers. Operators controlled most of the activities, i.e. operation and access, the sales and after-sales of handset and services, and end-userbilling (Maitland, Bauer, & Westerveld, 2002; *Roland Berger*, 2005). As such, operators controlled both the wholesale side of the value chain (i.e. the communications network) and the retail side (i.e. customer interaction and support services) (Sabat, 2002). The manufacturers developed network and device equipment. Although diversification and expansion strategies of both operators and manufacturers gave the market a seemingly dynamic look, the division of roles remained relatively stable.

The introduction of mobile data services disrupted this well-organized value chain. The high investments in UMTS licenses and the costs made to implement the diversification and expansion strategies led to financial problems for operators (Maitland et al., 2002). At the same time, the subscriber base for voice services was saturated, while the average revenues per user were falling (Olla & Patel, 2002). Data services were seen as the best way to recoup license and network roll-out investments, while at the same time increasing ARPU rates. Data services added a number of activities to the value chain, such as Internet access and development and provisioning of middleware, content, applications, platforms and portals (Jaokar & Fish, 2004; Maitland et al., 2002; *Roland Berger*). As a result, actors were entering the mobile services domain from the content, IT and consumer electronics sectors (Olla & Patel, 2002). Because it is no longer possible to identify the roles involved clearly, and many relationships between the actors go beyond the structure of a linear chain, it makes more sense to see the mobile data services industry as a value network (Li & Whalley, 2002).

In this section, we first discuss the roles and resources that are typically required for mobile data services. After that, we describe typical governance models used in these value networks by contrasting walled garden and open models of collaboration.

4.4.1 Roles in Mobile Value Networks

Actors providing mobile services can fulfill a variety of roles, and there is no generic role division that applies to all services. In this paragraph, we link the resources and capabilities needed to deliver mobile data services to the typical roles suggested in the STOF model.

Network Roles

For a service to be delivered, connectivity is required between the end-user and the service provider. This connectivity is provided by the network operator, i.e. the actor operating the cellular or short-range network. The *UMTS Forum* (2002) defines the key function of a network operator as the provision of access and transport services, typically holding the license to use the network. As network operators provide connectivity to the end-user, they have access to the customer. They also dictate which mark-up language content providers should use in delivering their content. The largest European operators are Deutsche Telecom (T-Mobile), Vodafone, France Telecom (Orange), Telefonica, Telecom Italia and British Telecom (O2). Generally speaking, the series of consolidation moves and expansions have led to a fairly consolidated market (in the USA there were nine operators before 2005, with only four remaining in 2006) (Estenfeld, 2006). Similar trends can be observed in individual European countries.

Mobile Virtual Network Operators (MVNOs) play the role of agent between the network operator and end customers (*UMTS Forum*, 2002). Kuo and Yu (2006) divide MVNOs into two categories; other telecommunication operators and non-telecommunication operators. The non-telecommunication operators are usually actors operating in different industries in which they already have an established brand and retail channel. Kiiski (2006) distinguishes data-only, voice only and voice and data MVNOs. As MVNOs purchase capacity from an infrastructure company, they have limited physical assets, and they will be able to use their resources in a different way should their core business become unsustainable (Li & Whalley, 2002). This makes them relatively flexible in terms of their service portfolio and relationships with other actors and allows them to focus on niche markets where possible. The number of MVNOs varies per country, e.g. in the USA, UK, Finland and the Netherlands there are many of them, while in Japan, Italy and South Korea none are active as yet (Netsize, 2007).

Hardware Roles

The network infrastructure consisting of antennas, base stations and core network is provided by network manufacturers. While they may not play a direct role in every single mobile service offering, they are important in the long term because they have a say in which new technological standards will be developed. Typically, a network infrastructure is leased to operators. Consolidation has taken place in the network manufacturing market as well, and there are currently four dominant players, i.e. Ericsson, Huawei, Lucent-Alcatel and Nokia-Siemens (Netsize, 2007).

End-users require handsets to receive services. These handsets are provided by handset providers. Especially for more advanced mobile Internet services, handsets are important. Network operators often bundle subscription packages with 'free' handsets in order to encourage the adoption of advanced handsets. In the current market, there are about 100 vendors producing over 2,000 different phones. The market is dominated by four major players, i.e. Nokia, Motorola, Samsung and Sony-Ericsson (Netsize, 2007).

User Related Roles

If end-users have to pay for a service, billing facilities are another necessary resource. Billing consists of several activities, i.e. charging, billing and accounting (Koutsopoulou, Kaloxylos, & Alonistioti, 2004). The billing and collections provider 'issues bills (or the equivalent) and arranges for collection of payments from customers' (*UMTS Forum*, 2002, p. 3), either through prepaid or post-paid billing arrangements. Accounting and dividing the revenues among the actors involved in offering the service is usually carried out by the same actor. The billing provider has a direct contact with the end-user. At the moment, it is usually the network operator who plays the role of billing provider, although other actors in the fixed Internet and banking domain are also adopting this role.

To allow for personalized services to be provided, users must be identified and authenticated. Typically, this is done by the same actor playing the role of network operator, by authenticating the user based on the SIM card. However, alternative models exist in which authentication can take place, for example, via web services.

In addition to demographic user information, information about the handset, browser and application on the handset is often required for a proper adaptation of the content to ensure a good service experience. This information can also be provided by the network operator, who typically has specialized systems to retrieve this type of information. However, alternative models exist, for instance by adding an application to the handset that transmits the relevant information.

Customer support is another resource that is often required for mobile Internet services. While many actors can be involved in offering a service, there is often one point of contact available to the end-user in case of problems. This role can also be played by network operators, or by content providers.

To allow context-aware services to be provided, dynamic information about the context of the user is needed. An important example of this type of information is location information, which can be provided by the network operator by triangulating the position of the user, but it can also be provided by an independent party, for example using GPS technology.

Software Roles

Most types of services require applications and platforms, both to run on user devices and at the service providers' end. Specifically, content adaptation platforms are often required to adapt web-based content to a WAP-based format that suits the specific user device. Users may require micro-browsers and Java applications to access services. Developments like MWS and, increasingly, intelligent devices pave the way for a more prominent role of software related actors in the near future.

Applications run on platforms that allow consumers to access Internet services on their mobile devices and enable enterprises to extend their commercial applications to the mobile network (Kuo & Yu, 2006). Because at the moment only few platforms are developed in-house, the actors involved primarily come from the computing industry, for instance operating system and middleware vendors (Tilson & Lyytinen, 2006).

To provide location-based services, expertise and data is needed on geographical information systems (GIS). For example, maps and geo-spatial data needs to be available to be able to offer routing information to users, and geospatial calculation will be required to compute meaningful spatial information from raw coordinates about the user position.

Content Roles

With regard to content services, various additional roles can be distinguished. First of all, raw content needs to be created. Kuo and Yu (2006) use the term 'content developers' to denote actors who provide, design and produce various kinds of products or services for all kinds of end-users. Raw content is typically sold on a wholesale basis to operators or providers of content and applications. As we discussed earlier, there is a trend towards user-generated content, and users can to a certain extent take over the role of the traditional content providers. However, content developers are also developing new formats based on this trend.

A content provider is an actor providing and distributing content (Grover & Saeed, 2003). Although content providers can create their own content, they usually provide content obtained elsewhere. The *UMTS Forum* (2002, p. 3) defines 'content provider as a provider of services that add value to access and transport services. Value-added services can be produced by the content provider itself or purchased from others.'

While content developers and providers may focus specifically on mobile services, for many organizations the mobile channel is just another channel to distribute their content. For example, broadcasting organizations, traffic information providers, banks and tourist information offices already provide content services to customers using traditional channels like TV or the fixed Internet.

Advertisers form a specific type of content providers, because they offer sponsored content. They basically provide sponsored content to content providers, aggregators and service providers, to be included in their content services, for which the advertiser pays a fee. Advertisers can also provide free (sponsored) services or content on their own. Advertisers can be any actors willing to pay for the distribution of his brand name and products and services. Advertising agencies will play an important role in delivering and possibly adapting advertisements to the mobile devices. Similar to the content providers, advertisers can focus specifically on the mobile channel or extend their marketing mix with mobile advertising.

When content from a number of providers is combined in a single service offering, a content aggregator is required. Generally speaking, there are passive aggregators, who merely bundle the content, and active aggregators, who carry out filtering, editing, or customization (Barnes, 2002; Li & Whalley, 2002).

As the amount of content on the mobile Internet is increasing, search engines become more important. Search engine providers can help manage content and the complexity of content. Often, actors from the fixed Internet, such as Google or Yahoo!, play the role of search engine provider.

As discussed in this chapter, many mobile Internet services are based on portals: 'a network site that aggregates, presents, navigates, and delivers a wide range of Internet communication, commerce, and content services to a large number of visitors.' (Sabat, 2002, p. 522). The portal is a 'gate' to the mobile Internet, as it is the first point of browsing (Kuo & Yu, 2006). Barnes (2002) places the portal under 'market making', as the portal is aimed primarily at marketing and selling content, including program development, service delivery and customer care. Compared to regular Internet portals, mobile Internet portals have a strong need for customization and personalization (Barnes). On the mobile Internet the most widely used portals are those of the operators. However, other content aggregators, for instance Internet players, increasingly provide mobile portals.

To assure the adoption of the service, promotion and marketing are often required. In principal, any actor involved in the service offering can play this role. Because operators have strong brand names and large marketing budgets, they often play this role.

4.4.2 Governance Models

Typically, network operators play many of the roles discussed in the previous section. As a result, content providers depend on operators to gain access to the customer, and to perform the activities involved in authentication, billing and localization. Operators have adopted various strategies to determine how much freedom they grant content and service providers in terms of access to their customers. Feijóo, Marín, Martín, and Rojo (2006) have identified a continuum between full operator control without participation from other parties, and full participation of others without operator control. These two extreme options are called the walled garden and the open model, respectively.

Walled Gardens

Jaokar and Fish (2004) define the walled garden as 'a mechanism to restrict the user to a defined environment, i.e. forcing them by some means to stay within the confines of a digital space'. In this model, users can only access content on the operator's portal, and have no access to other WAP sites (*OVUM*, 2006). Although most content on a portal is developed by third party content providers, operators take care of most activities including controlling, approving, and managing quality of the content, and navigation and customer care (Feijóo et al., 2006; Forrester, 2006b; *OVUM*). The brands of the operator and the content providers are more or less merged, and their relationship is close (*OVUM*).

For end-users, the advantage of the walled garden portal is that they receive a consistent end-user experience, because all content has the same look and feel (Forrester, 2006b). However, their freedom of choice is limited by the amount of content on the portal (Feijóo et al., 2006; Forrester). Operators choose walled garden models because they guarantee them large shares of the revenues (Feijóo et al.), and reduces the risk of them becoming mere connectivity providers. However, some analysts claim that operators, due to their limited creativity, have not appeared to be able to provide good user experience (Forrester). In addition, the visibility of operator portals is low, because operators mostly have to take care of promotional activities by themselves (Forrester).

The walled garden models are threatened by emerging billing solutions that bypass the operator, the increasing availability of good content, alternative solutions like PC-side loading to downloading of premium contents, and the increasing popularity of independent WAP sites (*OVUM*, 2006). In addition, operators are becoming increasingly aware of the strategic option of generating revenues with off-portal data traffic and additional revenues like sponsored search (*OVUM*). As a result, walled gardens are being opened up.

Open Models

In an open access model, users can access content from any provider (Feijóo et al., 2006). Aside from optional agreements on billing and authentication, there is no relationship between operator and content providers. In an open access model, customers can access more diverse content and content providers are not constrained by demands from the operators (Forrester, 2006b). Competition between content providers and greater freedom to experiment may lead to more innovative services. On the other hand, operators may stop investing in innovative network and middleware technologies if they are uncertain whether they can still recoup a sufficient portion of the revenues. Also, a lack of central billing, security and customer support may increase complexity for consumers (Forrester, 2006b).

Hybrid Models

There are several hybrid models for the coexistence of third party providers and content services by the operator, (Feijóo et al., 2006). In these models, operator and third party branded content are mixed. An example of such a model is the i-mode ecosystem model, which offers users official and unofficial sites that both use operator billing facilities, i.e. a semi-walled garden model that allows users to browse and optionally download content off-portal from a set of exclusively approved partners (*OVUM*, 2006). In this model, operators only provide connectivity and billing, and do not control the content provider using its billing services. Nevertheless, operators impose certain conditions on content providers, for instance a maximum content fee for end-users. A related model is the one used by Orange's Gallery WAP site, which provides access to other WAP sites to users from any network (*OVUM*).

Depending on the intended offering, a diverse range of resources and capabilities is required to provide mobile services. Actors who play roles relating to content, software, hardware and networks, have to work together to make the business model work. While governance mechanisms have traditionally relied on operators organizing the activities in a walled garden model, technological and strategic developments have put this position under pressure. As a result, the industry is shifting from closed models of collaboration towards more open models that allow flexible collaboration between content providers, application providers and operators.

4.5 Financial Arrangements

An equitable division of costs, revenues and investments is required to make the collaboration worthwhile for all the organizations in a value network. In addition, a proper end-user tariff has to be installed to collect the necessary revenues. In this section, we discuss models for the distribution of costs, revenues and risks among the actors in a network. After that, we discuss typical choices in end-user tariffs used to charge for mobile services.

4.5.1 Division of Costs, Investments and Revenues

There is a diversity of investment models for the various mobile industry actors. As far as content providers are concerned, most investments are made during the development of content and the costs of reproduction are typically low (Shapiro & Varian, 1999). To increase return on investment, cross-media strategies are often applied, i.e. the content is distributed to end-users over various channels, of which the mobile Internet channel is but one. It is generally difficult to predict the eventual market success of specific content. Often, investment is required in developing specific applications that allow the content to be offered to mobile users. While offering services, the main costs facing content providers include marketing and promotion, customer care, and possibly a fee to network operators for using their networks.

The main investment costs of operators have to do with building a network rather than with developing individual services. Capital Expenditure (CAPEX) comprises the costs of buying licenses and equipment, the installation of base stations and site engineering. While the costs of rolling out a cellular network are typically high, for short-range technologies such as WiFi the investment costs may be lower. Once a network has been rolled out, the marginal costs of adding users or content services is low. This explains the wholesale strategies of operators, in which excess network capacity is sold to virtual network operators of content providers. With regard to operators, operational costs involve leasing the site where base stations are installed, interconnection and roaming fees, electricity consumption, backhaul transmission, and operation and maintenance (Rendón, Kuhlmann, & Aranis, 2007). In addition, non-network related operational costs include service support, marketing and promotion, billing, bad debt and fraud, subsidies on handsets and customer care.

Revenues are typically collected by the operator who has a billing relationship with the end-user. To divide the revenues, a revenue share model is often used. For example, in the i-mode model the operator shares 84% of the revenues with its content providers. When using SMS-based charging for WAP services, the revenue share or kick-back fee may be reduced to 50%. In some cases, operators buy content from content providers in order to sell the content via the portal. Mostly, the operator pays a fixed fee to the content provider for this. Application providers often charge a fixed project development fee to content providers and operators for the development of the application, in addition to a monthly fee for support and maintenance.

4.5.2 End-User Tariffs

End-users typically pay for two service components: the use of the radio network resources (i.e. connectivity) and the value of content services delivered over that network (i.e. content). For connectivity charging, many models are being used, e.g., metered charging, packet charging, expected capacity charging, edge pricing, paris-metro charging, and market-based reservation charging (Cushnie, Hutchison, & Oliver, 2000). Basically, these models can be divided into fixed and packet-based charging. Fixed or flat fee charging is based on a fixed fee regardless of the actual traffic on the network, while packet-based charging is based on the amount of traffic generated or time spent online. While packet metering used to be the dominant way of charging for network resource usage, flat fees are beginning to enter the market. The amounts charged for network resource usage vary. The highest tariffs are found in Japan and Germany, followed by the USA, and they are lowest in South Korea, Denmark and Finland (*TNO*,

2006). Generally speaking, the average revenues per user (ARPU) are decreasing, especially when it comes to services (*Roland Berger*, 2005). A large proportion of these revenues is still related to voice services, while data ARPU currently comes mainly from plain SMS services rather than from advanced mobile data services (Forrester, 2006b).

While customers are commonly used to paying for connectivity, they are less used to paying for content, because on the fixed Internet content is often free. For content charging, various models are in use, e.g., pay-per-download, subscription based, event based, context pricing, personal pricing and value-based pricing. In terms of technology, various models are used for content charging, such as premium-SMS and WAP billing. Other models for generating revenues, such as advertising and sponsoring, are increasingly considered.

Generally speaking, there are two types of tariff schemes. In post-paid billing, users generally pay a fixed fee per month as well as a fee for calling and data traffic. Often, users get their device for free with the subscription, through subsidization by the operator. In the case of prepaid billing, users pay in advance and calls and data traffic are credited against the prepaid balance. This scheme is especially common at the lower end of the market. While prepaid and post-paid tariffs used to be strictly separated in the past, they are currently converging, because users may want to pay for some services through a pre-paid system (e.g. content downloads), while they want to be charged for more predictable costs (e.g. monthly contract or voice calls) via a post-paid billing system (*AtosOrigin*, 2006).

Generating revenues with mobile services is challenging, as customers are not used to paying for content services on the fixed Internet. The industry is trying to solve this through flat fee models for network resource usage, and a mix of flat fee, fee per usage and free, advertisement based models for content services. Typically, revenues are collected by operators that have the billing relation with the end-user, and are divided among content providers through revenue share models.

4.6 Typical Mobile Services Business Models

Thus far, in this chapter we have illustrated the context in which mobile services business models are developed, and how this context is different from that of other e-services. While there are various largely accepted classifications of fixed Internet business models (e.g., Rappa, 2000), their applicability in the mobile domain is limited (Leem, Suh & Kim, 2004). Several authors propose alternative mobile business model classifications or atomic business models. Typical shortcomings of these models include their focus on the service component rather than the full business model and their bias towards consumer-oriented rather than business-oriented models (Leem et al.). In this section, we discuss classifications based on generic principles, design parameters, and dominance of players. Because we focus on business models on the level of the entire value network, role level classifications such as the one proposed by Camponovo and Pigneur (2003) fall outside the scope of this chapter.

A common classification of business models is related to the allocation of roles to the various actors. Panagiotakis, Koutsopoulou, and Alonistioti (2005) and *UMTS Forum* (2002), for example, discuss three basic business models based on who is the dominant player, i.e. the network operator centric model, in which the operator owns the customer contact and performs roles of connectivity provider, content aggregation and charging; the service/content aggregator centric model, in which the content aggregator combines content from third parties into a service offering and collects payment independent from the network operator; and the service/content provider centric model, in which the publishing of content and billing/payment is conducted by the content provider. A key parameter in these three models is customer ownership, which is located respectively at the operator, content aggregator and service/content provider. An example of applying this type of classification to mobile payment is provided by Van Bossuyt and Van Hove (2007), who distinguish carrier-centric and payment service provider-centric models. A related discussion regarding business models for mobile Internet services is the contrast between Internet-based and telecommunications-based business models. In the telecommunications business model, access to the network and access to content are controlled by the same entity, i.e. the operator, whereas in the Internet model these roles are divided, because Internet service providers typically do not provide content services. A clash between these models can be seen in the mobile Internet domain.

Killström et al. (2006) distinguish four generic business models based on innovative mobile services in the consumer market, by identifying the most important component of the business model. In the technology-driven model, technology providers play a central role; the advertising-based business model focuses on an advertising based financial component; the mobile extension model focuses on extending existing business with a wireless component; and their fourth business model focuses on selling content through the mobile channel. Methlie and Pedersen (2007) distinguish three dimensions in mobile business models: service strategy, i.e. mobile specificity, breadth of proposition, and market focus; governance form, i.e. hierarchical (closed) and relational/market (open);

and revenue model, i.e. content-based and transport-based. They focus on the contrast between open (e.g., Nordic countries) and closed (e.g., i-mode) models of collaboration, and their unclear impact on performance. They argue that internal choices influence the extrinsic service attributes of indirect and direct network effects and intrinsic service attributes of ease of use, usefulness, compatibility, service quality and innovativeness. They theorize, for example, that hierarchical forms lead to a lack of diversity in services, but a higher quality when it comes to complementary services. And that relational governance is positively related to innovation speed.

Ballon (2007) argues that a classification of mobile business models should be based on a set of key design parameters and a limited set of options per parameter. He proposes four levels of mobile services business models in which design parameters can be found, thus adapting our STOF model. These levels are the value network level, comprising the specific combination of assets, the level of vertical integration, and customer ownership; functional model level, comprising the modules and interfaces between models, distribution of intelligence within the system, and the interoperability with other systems; financial model level, comprising the cost (sharing) model, revenue model, and revenue sharing model; and value proposition level, comprising positioning, user involvement and intended value. For each of these parameters, he identifies the two main options that require a trade-off and that can be used to classify mobile business models. The approach is applied in the classification of business models for flexible spectrum provisioning by Delaere and Ballon (2007), and in the analysis of peer-to-peer versus IMS business models by Braet and Ballon (2007).

For each design parameter, Ballon provides only two extreme options. For example, his design parameter *Intended Value* is either 'price/quality' or 'lock-in'. He argues that any other approach would produce an almost infinite number of potential mobile business models. However, it could be argued that at least some of the parameters actually offer a continuous range of options. To stay with the example of *Intended Value*, one could imagine that companies chose to offer both a good price/quality ratio and at the same time try to lock-in their customers, rather than choosing one of the extremes. Moreover, some of the parameters could have options on multiple dimensions. *Intended Value* could have other dimensions, such as reliability, flexibility, ease of use and security, depending on the application domain, as Ballon (2007) argues. As such, the classification constrains the richness of potential mobile business models under study, and may even lead to false dilemmas. In addition, Ballon's approach is inconclusive about the causal relationships and trade-offs between the choices, i.e. whether or not they are expected to be interdependent.

Moreover, the approach is high-level and does not provide practical guidelines on how to design and develop mobile business models. In sum, although Ballon's approach appears useful when it comes to analyzing mobile business models on a high level of abstraction, it is not applicable as a design approach. In the next chapter, we provide our own approach to mobile business model design which provides more practical guidelines.

Chapter 5 The STOF Method

H. De Vos and T. Haaker

The STOF model describes the main concepts and design variables within the four business model domains, i.e. service, technology, organization, and finance (see Chap. 2). We have argued that the design choices that are made in these domains need to be balanced to realize viability and feasibility. To support a balanced design we introduced Critical Design Issues (CDIs) and Critical Success Factors (CSFs), as well as the specific issues regarding the business models of mobile services (see Chaps. 3 and 4). In this chapter, we introduce the STOF method, using a step-by-step approach to create a business model design for a specific service concept (Bouwman, Haaker, & De Vos, 2005; Bouwman, Haaker, Steen, & De Vos, 2003; Haaker, Oerlemans, Steen, & De Vos, 2004). The STOF method explicitly helps designers to create viable, feasible and robust business models that create value for customers and providers alike.

The method is especially useful in the early stages of service innovations: the exploration and initial elaboration of various ideas and options. When the method is used at an early stage, the service and technology design can be adjusted to satisfy the business requirements and increase market potential at a later stage.[1]

The STOF method consists of four steps. In the first step – the Quick Scan – an initial sketch of the business model is made. This basically means a description of the service and the intended value proposition, a value network, a technical architecture and a financial model. In the second step – evaluation with CSFs – the viability of the Quick Scan result is assessed. The initial business model is refined in step three, by specifying the CDIs. A robustness check is carried out in step four. These steps will be further explained in the next sections.

---

[1] To facilitate the use of STOF method, a handbook is available via http://www. stofmethod.com, providing templates for the design of business models using the four-step method. It can be adjusted to suit the specific needs and issues of a design session and is available in Dutch as well as English.

The STOF method can be applied in different ways. For example, individual designers can use it to structure and guide their design and design process. An especially fruitful way of using the STOF method is through a design session, i.e. a workshop in which multiple participants discuss the business model issues as outlined in the method and together formulate design choices. We discuss different ways of applying the STOF method in the final sections of this chapter. First, we turn to the design process.

Illustrative Example

Throughout the description of the STOF method, we use one example to illustrate the various aspects of the design. The example is based on a new mobile service idea for which design sessions were conducted including groups of researchers during a workshop about business models and business model design. These sessions were organized while the STOF method was being developed, and were among the first occasions that the method was being used. The service involves a guided tour along the Dutch megalithic tombs – so-called Hunebeds – with the use of a mobile device. Examples of such mobile tourist guides are currently commercially available, for example Abel (2007), which provides personalized cycle tours from hotel to hotel. The services Abel provides operate on a dedicated PDA without mobile connectivity. Similar services are the mobile navigation services of, for example, TomTom and Garmin, which operate locally on a (dedicated) device, with optional real-time support services and add-ons.

5.1 Design Process

The STOF method consists of four subsequent steps, illustrated in Fig. 5.1. The input that is required is an initial new service idea or service concept. Step 1 refers to the Quick Scan, in which specific design variables of the four domains are examined and initial design choices are formulated for the service idea under investigation. Answers to basic questions regarding the service concept, the technological architecture, and the organizational and financial arrangements, yield a broad outline of the business model. The business model concepts considered in the Quick Scan are related to the domain models of Chap. 2.

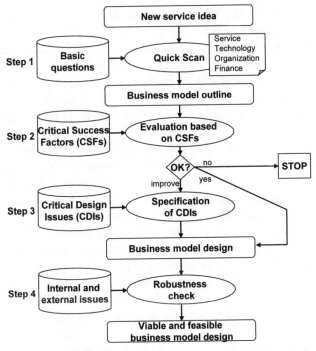

Fig. 5.1. Design steps in the STOF method

In step 2, this outline is evaluated on the basis of the eight CSFs (see Chap. 3), with the aim of assessing the expected viability of the business model. One example of such a critical success factor is the extent to which the proposed value proposition appeals to the target group. The evaluation helps to determine which parts of the business model have to be modified. Alternatively the business model is fine and further specification is not necessary, or the evaluation step shows that no viability can be achieved and the design process is stopped.

If there are doubts with regard to certain criteria, specific elements of the business model are reexamined or further specified in step 3. Here, the relevant basic information provided by the Quick Scan is worked out in greater detail and the business model is refined using the CDIs that were introduced in Chap. 3. The CDIs are related to design choices that have a strong influence on the assessment of the CSFs. These choices typically demand a careful balancing of the requirements and/or the interests of the various parties involved. The design issue *Personalization*, for instance, not only affects the attractiveness of the value proposition, it also has an impact on the costs of the service and on the requirements concerning the management of user profiles. Both refining the elements of the business model and balancing the interests lead to modifications in the original business model design. Steps 2 and 3 can be repeated in an iterative process, until a feasible and viable business design is created, or a decision can be made to stop the process due to a lack of business potential. While this may seem an unsatisfactory result at face value, it is in fact a very efficient way to filter out less promising service ideas at an early stage. Steps 1 to 3 have an internal focus, i.e. the design of a specific service with added value for customers and existing partners. Step 4 involves an internal evaluation, i.e. checking the relationships between the domains, as well as an external evaluation, i.e. focusing on the robustness of the design, for instance the sensitivity of the business model with regard to changes in the value network (What if new partners join? What if partners leave?), or its sensitivity with respect to competition.

We want to emphasize that in this chapter we provide the basic outline of the STOF method and describe the most important steps and issues. The workbook (see Footnote 1) is more refined and provides greater detail. After the four steps have been completed, a business model has been designed that, in the opinion of the designers, is viable and feasible with respect to the context and conditions that are expected. Of course, this is no guarantee for a successful service design. The advantage of using a systematic approach is that it provides a more rigorous business design, thereby reducing the chance of overlooking important issues and prevent market failure. Based on the business model that emerges, follow-up activities can be formulated and a start can be made with the development of an actual business plan or business implementation (cf. Chap. 6).

5.1.1 Step 1: Quick Scan

In the Quick Scan phase, a rudimentary business model is developed, which includes aspects from the four domains of the STOF model. The Quick Scan focuses on design choices for design variables in the descriptive domain models presented in Chap. 2. The Quick Scan typically starts with the service design.

Service Design

The Quick Scan starts by outlining the value proposition from the point of view of the customers and/or end-users. We need to keep in mind that the customer of a service, i.e., the one paying for the service, is not necessarily also the user of the service. The main issue with regard to the service design is, therefore, 'What is the service concept, and what is the added value for the customer and/or user?'

To answer this question, service design focuses on the following set of design variables (cf. Chap. 2): *Intended Value* of the service, *Customer* and *End-user, Context* of use, service *Tariff* and *Effort*. Typical questions that need to be answered are 'Who is the customer, who is the user?', 'What is the specific service? Why would someone want to use it?', 'What is the context of use?', 'Are there alternative products or services?' and 'What would customers be prepared to pay?' The result is a clear description of the service, its target customers and the intended added value for these customers, sometimes illustrated by a 'walk through' (see Fig. 5.2). Based on these results, requirements and restrictions with regard to the technology, organization and finance domains can be identified.

The ideas, assumptions and requirements that are addressed in the service domain serve as a starting point for the other domains. It is our experience that, when addressing other domains, it is often necessary to reconsider design choices that have been made in the service domain.

The Hunebed Case: Service Design

A regional Tourist Board wants to offer a personalized tour along the Dutch megalithic tombs – the so-called Hunebeds – in the region. Broadly speaking, there are two groups of tourists that visit the region: families with (young) children, who visit during the holiday season, and active seniors, who come on short breaks. Because the interests of these groups are quite different, the service will be designed specifically with the senior target group in mind. Because most senior couples visit the region by car, the service should support car-based mobility. The location-based service is provided via a dedicated PDA, equipped with GPS and small speakers. The PDA can be rented and dropped off at any tourist office in the region. It can be powered using the car's cigarette lighter. All information is stored on the PDA. The GPS is used to determine the customers' location and to ensure that relevant information is provided. The service provides navigation and tourist information based on the location of the customers.

Alternatives to this service are traditional paper guides. The added value compared to paper guides is that the information that is provided is location-specific and the service offers various multimedia options. Generally speaking, paper guides costs around € 5 to € 6, which means that the consumer price for the new service should cost about the same. The service is called Hunebed&Breakfast, since it is targeted at people who take short breaks, and also provides information on restaurants, hotels, et cetera. An illustration of the service concept is presented in Fig. 5.2.

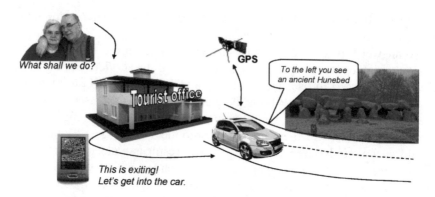

Fig. 5.2. Service design walk through for the Hunebed case

Technology Design

To deliver a value proposition requires a variety of infrastructure and technologies. Following the core variables of the STOF model (see Chap. 2), the technology design outlines the design choices for *Technological Architecture*, which consist of *Applications, Devices, Service Platforms* (including *Billing* and *Customer Data Management*), *Access Networks* and a *Backbone Infrastructure*. Furthermore, the *Technical Functionality* is defined and the specific *Data* streams are specified.

The technology domain focuses on the core technologies that are required, and on their possibilities and limitations. The technologies that are selected should match the results of the service domain. Technology should be in line with the value proposition, be usable for and available to the customers and match the expected price level expressed in the service domain. Often, there are various ways to realize a value proposition. With regard to high-end services with quality as a core aspect, an expensive solution is suitable. In a mass market that is characterized by low customer fees, a lean and less expensive solution is preferred. The value network should be aligned with the technology design. In many cases, iterations between designs in the technology and organizational domains are needed to realize this alignment, for example when existing hardware and software have to be incorporated into the value network.

The main questions in the technology domain are 'Which technologies are needed?', 'What are the specific requirements?' and 'How can the infrastructure be specified?' Based on the answers to these questions, an initial estimate of the costs can also be made.

The Hunebed Case: Technology Design

The technology that will be used in the Hunebed case has been roughly defined in the service domain, since it is very much a part of the value proposition. A decision is made to use PDA's. As far as the intended target group is concerned, it is important that the application be easy to use, and the content and symbols should be clearly legible. A low-cost solution is selected, with all the content stored on the PDA and periodically updated by the Tourist Board with aggregated information based on the content of several content providers. According to the service design, all information is stored on the PDA, which means mobile connectivity is not required. Obviously, this is a very important design choice. If real-time data transfer was required, the technology, organization and finance design would be quite different. The solution would become much more expensive. Instead, all the maps, content and software are stored on the PDA, and can be periodically updated.

Since most customers travel in pairs, a PDA with built-in speakers makes it possible for both persons to listen to the audio. The information is linked to the customers' location. A GPS module is used to determine their location. A possible technical architecture is presented in Fig. 5.3. The application can be easily modified for use in other regions, simply by changing the content.

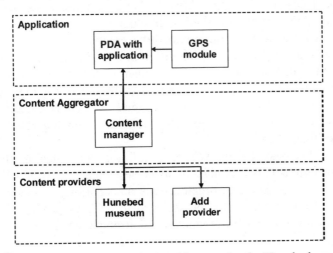

Fig. 5.3. High-level technical architecture for the Hunebed case

Organization Design

After defining the service concept, the value proposition and the technological infrastructure, the organizations that have to be involved in delivering this service concept can be selected. In most cases, multiple parties are needed, since mobile services require access to complementary assets provided by various organizations. For a co-operation to be sustainable, every organization must benefit, either in tangible or in intangible terms. Furthermore, the goals and interests of the partners must be aligned, and not clash. Following the core variables of the STOF model (see Chap. 2), the organizational design focuses on the *Activities* that combine into *roles*, the *actors* that have the required *Resources and Capabilities* to play these roles, and the *Organizational Arrangements*.

To begin with, a list of roles can be drawn up with regard to the activities and resources needed to deliver the service. Each role can be characterized by the output it provides in terms of service, product and value. The roles, together with the value exchanges between them, determine the *Value Network* of the service. The exchanges that take place may be of a tangible (providing devices or content) or intangible nature (providing financial or marketing-related expertise). Each role has specific responsibilities and will deliver a specific resource or capability needed by others. It is useful to identify the structural, integrative and the supporting, facilitating roles in the value network. What specific roles the various actors play will depend on their individual resources, capabilities and strategic interests, and on the competitive environment. Often, multiple value networks are possible, which differentiate with respect to the actors that take on structural roles. The Quick Scan of the Organization domain results in a preliminary design for the actors involved, the organizational arrangements, and the value network structure, as well as a description of the actors' strategies, goals and value activities, and the resources and capabilities they contribute.

In some business design sessions, defining the roles in the value network design will be enough. In other sessions, it is useful to assign specific actors, for example organizations or departments, to the various roles. Because roles can often be assigned to different actors, in the end this leads to several possible organizational arrangements.

The Hunebed Case: Organization Design

Normally speaking, the Tourist Board would be the initiator of the service as well as the central actor in the value network (see Fig. 5.4). Information providers (or content providers) are needed to provide information regarding regional points of interest. The Tourist Board takes care of the physical distribution of the PDAs, the marketing and the aggregation of the content. The Tourist Board takes care of customer contacts. To develop and maintain the PDA applications, the services of an application developer are used. A device provider will handle the supply and maintenance of the PDAs. Finally, the Tourist Board has to negotiate agreements with advertisers. The Tourist Board may cooperate with local shops and tourist offices to ensure customers to be able to pick up and to return the PDA's at any address. Figure 5.4 provides an illustration of the value network.

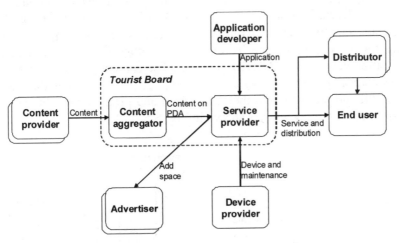

Fig. 5.4. Hunebed case value network

Finance Design

The bottom line of the business model is addressed in the finance domain. According to the STOF model, this involves the following design variables (see Chap. 2): *Revenues* on the one side and *Investments*, *Costs* and *Risks* on the other. The main issue is whether the service yields sufficient revenues for all the parties involved to be compensated for their efforts, costs and risks? First of all, the finance design outlines the *revenue sources* and *the cost-, investment* and *risk sources*. Taking part in a value network can have other, less tangible, benefits, in addition to monetary revenues, for instance market expertise, access to a new technology or insight into the business strategies of relevant business partners. With regard to the viability of a business model, it is important to make sure that all the parties involved somehow benefit, which means that finance design outlines the *Financial Arrangements* that cover an acceptable *Division of Revenues*, investments, costs and risks. The financial arrangements in the Quick Scan will initially be based on the designs in the other domains and on assumptions regarding cost levels, tariffs, and number of users, et cetera.

The Hunebed Case: Finance Design

End-users will pay the Tourist Board a fixed daily rate of € 6. This price is set to match the price of the competitive paper guides. Making a profit is not a target for the Tourist Board, in fact, breaking even is enough. In addition, advertisers will pay the Tourist Board for placing their location-based ads on the PDA and directing customers to interesting offerings along the way. Finally, the Tourist Board settles the account with the application developer and device provider. The content providers think it is such a charming idea that they do not need to be paid. They feel that taking part in this project will help their reputation and promote their brand. The distribution of the revenues is shown in Fig. 5.5.

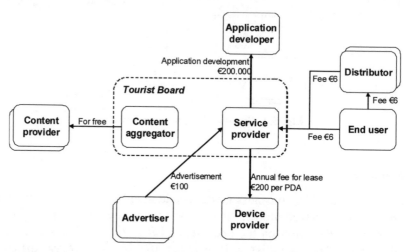

Fig. 5.5. Revenue streams in the value network for the Hunebed case

As far as the technology design is concerned, a relatively 'cheap' and basic solution is selected that does not involve any real-time data exchange. This decision was made for financial reasons: the solution was sufficient to deliver the value proposition at a reasonable price.

Table 5.1. Hunebed case's rough estimates of annual revenues, investments and costs

Revenues		
From usage, based on a € 6 fee per use and 15,000 customers	€	90,000
Advertisements, based on 100 advertisers paying € 100 each	€	10,000
Total revenues	€	100,000
Investments		
Development and maintenance application	€	200,000
Exploitation costs		
PDA lease, based on 100 PDA's and € 200 fee per PDA	€	20,000
Annual budget for marketing and communication	€	10,000
TOTAL annual exploitation costs	€	30,000
Annual result		
Revenues	€	100,000
Depreciation of investments (in 3 years)	€	66,667–
Exploitation costs	€	30,000–
Result	€	3,333

Rough financial estimates can be made to form an impression of the returns on investment. The estimates presented in Table 5.1 are based on the availability of 100 PDA's, which are expected to be used 150 days per year. Including maintenance, the leasing costs for the devices is about € 200 per year. The development of the application and maintenance will cost € 200,000 over a three-year period. Based on these assumptions, an annual positive result of a little over € 3,000 is expected. Provided the Tourist Board manages to realize the content aggregation and distribution for this amount, the project will break even.

Balancing the Domains

After the Quick Scan has been carried out, there are four domain-related designs that together make up the initial business model: a description of the service and the intended value proposition, a value network, a technical architecture and a financial model. It is likely that the requirements in the four domains affect each other. Before moving on to the next stage – evaluation – fine-tuning between the domains is relevant. Because the Quick Scan design is relatively sketchy, this can be done in a quick, relatively basic and qualitative way. Questions that drive the balancing activities between domains are listed in Table 5.2.

Table 5.2. Balancing the Quick Scan domains

	Service domain	Technology domain	Organization domain
Technology domain	Can the technology design deliver the value proposition?	–	–
Organization domain	Is the value network able to deliver the service and its value proposition?	Does the value network match the technology architecture?	–
Finance domain	Are service fees in line with the value proposition?	Are the estimated investment regarding implementation realistic?	Has a win-win situation been achieved?

To increase the balance between the domains, some simple adjustments can be made to the business model before proceeding to the evaluation. Additionally, the domain models can be checked superficially with regard to their robustness, simply by asking what-if questions like: 'What will happen if one of the partners ends the participation?', and 'What if a competitor offers a similar service?'

5.1.2 Step 2: Evaluation with CSFs

The evaluation of the Quick Scan design looks at how well the business model satisfies the CSFs for business model viability. The CSFs have to do with creating value for end-users (customers) and service providers, as discussed in Chap. 3. The evaluation is based on the causal models presented in Chap. 3, which state that design choices with regard to the CDIs have a strong influence on the CSFs, and thereby on business model viability as well. Consequently, the CDIs are of the utmost importance to business model designers and managers. By carefully specifying the CDIs they can positively influence the CSFs and as a result the business model's viability.

The evaluation in step 2 focuses on the CSFs. The underlying logic is that a negative assessment of certain CSFs implies that there will be bottlenecks in the business model's viability, and that CDIs related to such CSFs should be redesigned. Table 5.3 provides an overview of the CSFs and CDIs that should be reexamined when a CSF is evaluated negatively. For an elaborate description of CSFs and CDIs, we refer to Chap. 3.

Table 5.3. Overview of the CSFs and their relationship to the CDIs

Critical Success Factors	Related critical design issues in the factors			
	Service domain	Technology domain	Organization domain	Finance domain
Clearly defined target group	Targeting, creating value elements	Accessibility to the target group		
Quality of the value proposition	Creating value elements, branding, customer retention	Security, management of user profiles, accessibility to the target group		
Quality of the service delivery		Security, quality of the service, Management of user profiles, system integration		
Acceptable division of roles			Partner selection, network openness, network governance, network complexity	
Acceptable profitability				Pricing, division of investments, valuing contributions and benefits, division of revenues
Clear joint strategy			Partner selection, network openness, network complexity	
Unobtrusive customer retention	Customer retention	User profile management		
Acceptable risks				Valuing contributions and benefits, division of investments

The evaluation of the CSFs can be qualitative or quantitative in nature. For example, the Quick Scan designer(s) or others may express a general impression of the extent to which the CSFs are satisfied, or they may adopt a more formal approach to scoring the CSFs. Hence, the results of the evaluation in step 2 are a set of (qualitative) judgments on each of the CSFs identified in the Quick Scan design, rating them as either satisfactory or unsatisfactory. When at least one CSF is unsatisfactory, the CDIs related to this CSF have to be refined or re-evaluated. In addition to reconsidering the CDIs, it is also possible to add more detail to the core concepts.

The Hunebed Case: CSF Assessment

The assessment of the CSFs in the Hunebed case is summarized in Table 5.4.

Table 5.4. Selection of CSF assessments for the Hunebed case

CSF	Assessment
Clearly defined target group	*Positive.* The service focuses on short breaks for senior couples, a clearly defined target group
Quality of the value proposition	*Negative.* The service supports these short breaks by offering personalized tours and tourist information. It assumes that such a service will appeal to senior couples, which ideally should be confirmed, for example through market research. Another issue has to do with the distribution of the service, which is organized through the local tourist offices, with specific opening hours. It is likely that the target group requires a greater flexibility, such as more pick-up and drop-off points, independent of opening hours
Quality of the delivery system	*Negative.* To reduce the probability of a break-down, a robust device can be chosen. Since connectivity is not an issue, no problems are expected with regard to the technology. However, a helpdesk should be available to assist in case of any problems or questions
Acceptable division of roles	*Positive.* The tourist board orchestrates the network, contacts advertisers and is responsible for the distribution. Since the service is a regional initiative without profit goals, the division of roles is acceptable
Acceptable profitability	*Positive.* Advertising fees are low and advertisers expect additional revenues to outweigh their expenses. Profit is not an issue for this service, although breaking even is desirable
Joint network strategy	*Positive.* The network consists of parties in the tourist sector that focus on bringing visitors to their region and to specific sites of the network participants. The service offering is a new way to realize this objective, which fits a joint strategy

Quality of the Value Proposition and *Quality of the Delivery System* are unsatisfactory. If we look at Table 5.3, this implies that several CDIs from the service and technology domain, for instance *Creating Value Elements* and *Quality of the Service*, should be reconsidered.

5.1.3 Step 3: Specification of CDIs

In this step, a selected set of CDIs is specified in greater detail, depending on the evaluation of the CSFs in step 2. Generally speaking, there are two ways to approach this refinement. Since CDIs are related to the domains as well as the CSFs, one approach is simply to consider the CDIs for each of the domains, while the other involves refining sets of CDIs for each of the CSFs.

Refining CDIs for each domain starts with the service domain, for example by addressing *Targeting* or *Creating Value Elements*. This may lead to a redesigning or detailing of design variables like *Market Segment* or *Value Proposition*. After that, the next domain is examined, and so on. When all the CDIs have been refined, the domains have to be balanced. This process is visualized in Fig. 5.6. If necessary, some of the basic issues in the domains can also be reconsidered.

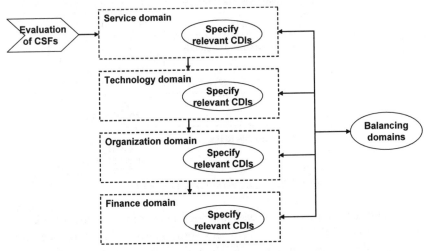

Fig. 5.6. Specification of CDIs per domain

The alternative approach involves re-evaluating sets of CDIs that have a strong influence on the CSFs, i.e., following the evaluation of Step 2. These sets consist of related CDIs in the various domains. For example, *Targeting* in the Service domain and *Accessibility* in the technology domain are both related to the CSF *Clearly Defined Target Group*'. Hence, when the target groups are further specified in the Service domain, the focus in the technology domain is on how these target groups are to be given access to the service: Which technologies do they use? Do they need to purchase specific equipments? The result will be a balanced technology design with specific target group characteristics. When refining CDIs for each CSF, attention is paid to balancing the domains automatically, due to related CDIs. This approach is illustrated in Fig. 5.7. Also, this approach will lead to redesigning or detailing design variables.

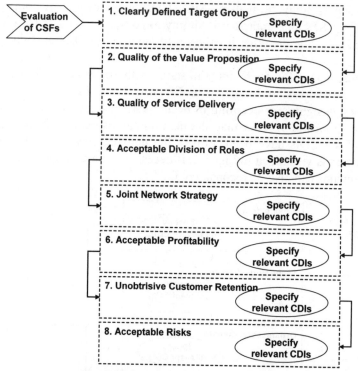

Fig. 5.7. Specification of CDIs per CSF

Refining a CDI involves a well-founded selection of a specific design option for that CDI. Chapter 3 provides some general clues with regard to the available choices for every CDI, and the consequences of specific design choices. In addition, the balancing requirements are also described, i.e. the relationships with other business model domains, and the criteria that guide the selection of a specific design option. Table 5.5 provides an example of the options, results, balancing requirements and criteria that apply to the CDI *Targeting*, which is linked to the success factor *Clearly Defined Target Group*.

Table 5.5. Aspects of CDI *Targeting*

Design issue: *Targeting*		
Options	1: generic service	2: aimed at a specific target group
Consequences	Mass service	Tailor-made solution
Balancing requirement	Fit between the set-up of primary processes (large-scale or small-scale, expertise areas), scale of services and wishes of end-users	
Criteria	Access to specific target groups	
	Size of company and primary processes (small company fits tailor-made solutions, large fits mass services)	
	Areas of expertise	

The example provided in Table 5.5 is described in general terms. The exact design options, consequences and criteria will of course be case dependent. Determining these options and other elements is an important part of the creative business model design process.

The Hunebed Case: Specification of CDIs

As far as the Hunebed case was concerned, the CSFs *Quality of Service Delivery* and *Quality of the Value Proposition* were unsatisfactory. This implies that the related CDIs should be refined or reconsidered. One of the CDIs involved is *Creating Value Elements*, which refers to value creation for the targeted users of the service, i.e. senior couples. Elements regarding mobile ICT services that are relevant to the Hunebed case are accuracy, support and distribution. Accuracy involves the exact positioning of the users and an up-to-date map that contains sightseeing information and small roads for hiking and cycling. This needs to be taken into account in service development in the technology domain. The service aimed at a target group that is not necessarily used to this kind of technology. A user-friendly device and software are necessary, including support in case of any questions and problems. The PDA could be equipped with a direct telephone connection to a helpdesk. Alternatively, a help function or frequently asked questions could be provided. Distribution is another value element. The initial business model uses the local tourist offices for distribution, which means users are dependent on a limited number of pick-up points, with limited opening hours. Preferably, every small village has at least one location where PDAs can be picked up and dropped off throughout the day. This requires a network of distributors, a well-designed PDA tracking and logistics system, and a way to reduce expenses due to loss, theft and damage.

5.1.4 Step 4: Evaluation

The focus of the business model design process has been on realizing a viable design, adding value for the intended customers as well as the providers. Step 4 involves the evaluation with respect to robustness and adaptivity. As we have seen in Chap. 2, business models evolve over time under the influence of the external business environment. Robustness of the business model has to do with the ability to cope with changes in the business environment. Is the service, for instance, dependent on the availability of complementary products or services which it does not control? In addition to robustness, the business model's capacity to adapt to external influences is an important evaluation criterion. Typical examples of external influences are changes in user requirements, regulatory changes, emerging new target groups and changing scale of operation, the application of a different revenue model or the incorporation of a new technology.

Robustness may be assessed by asking what-if questions, for instance:

- Is the design of the technological architecture modular? If this is the case, modifications to the technological infrastructure can be carried out on a modular basis.
- Is there a vendor lock-in that makes it hard to switch between suppliers?
- Is the technological architecture capable of absorbing a new and cheaper technology should such a technology become available on the market? And what are the consequences with regard to the other domains?
- What will happen to the organisational and technology domain designs should there be a huge demand for the service (100 times the estimated demand)? What are the consequences of such an event with regard to the other domains?
- What will happen to the Service design when it becomes clear that the service appeals to other target groups than expected, for instance retail customers rather than business customers?

5.2 Applying the STOF Method

In the previous sections, we have provided a fairly generic description of the 'what' of the STOF method. In short, the method consists of a structured approach to designing a business model for an idea or concept. The method includes questions and assignments that guide the designers through the various domains. In addition, criteria are provided for internal evaluation, i.e. whether or not the designers are satisfied with CSFs, and external evaluation, i.e. whether the business model is robust and can adapt to external influences. In this section, we take a look at *how* the answers to the various questions and assignments may be obtained, i.e., how the method can be applied.

The aim of the STOF method is to provide a structured approach to designing viable, feasible and robust business models for (mobile) services. As such, it is relatively unique, since many design methods contribute little to service definition and design (see Chap. 1). To the extent that similar approaches exist, they focus on project learning and time to market (Blazevic, Lievens, & Klein, 2003), issues that are only marginally related to our approach. For example, Quality Function Deployment (QFD) (Clausing, 1994) is not very suitable for mobile services, although an adjusted version aimed at mobile services is available (Herzwurm, Schockert, Breidung, & Dowie, 2002). System engineering lacks a multiple perspective approach. Van De Kar and Verbraeck (2007) have developed a system engineering method that is inspired by the STOF model.

Wohltorf and Albayrak (2005), describe how an investment decision support system may be used to assess and evaluate the robustness and viability of the business models of mobile services. Compared to our approach, their method is more telecom-oriented and less flexible in the specification of the design choices.

There are a number of ways to apply the STOF method. Depending on the design goals, different ways of designing and discussing the business model and of acquiring relevant business model data can be chosen. To obtain relevant market data desk research, interviews, surveys and focus groups can be used. For the discussion and actual specification of the design choices, facilitated design sessions and action research can be applied (see Table 5.6 for an overview).

5.2.1 Alternative Ways to Use the STOF Method

In Table 5.6, we provided an overview of common types of research with regard to applying the STOF method. However, when it comes to applying the method, there is no single success formula. Often, a variety of tools is used, depending on the specific goals of a given business model design. In this section, we discuss a few generic goals: gaining insight into the viability of a new service concept, designing alternative business models for a research prototype, and realizing a service design based on existing services.

Table 5.6. Overview of common ways of using the STOF method

Tools	Description	Typical use
Desk research	Searching, analyzing and interpreting literature	Market scan, case descriptions. Relevant to all domains. Typically applied in the Quick Scan or refinement step
Focus groups	Potential users discussing service concepts	Discussing service domain issues. Refinement step
Interviews	Obtain individual opinions and preferences	Applicable in all steps. Relevant to all domains. Obtaining input or feedback from experts, potential users, stakeholders, et cetera
Workshops	Discuss opinions, exchange of ideas	Applicable in all steps. Relevant to all domains. Alternative to interviews
Market research	Obtaining service domain input from future customers and users, e.g. by surveys	Research into added value and service domain issues in refinement step
User trials	Pilot or try-out by users of a demonstrator or prototype	Evaluation of value aspects and design choices. Typically this is a refinement step in the service and technology domain
Investment assessment	Assessment of risks, uncertainties and tangible and intangible benefits	Calculation of returns, investments and costs in the refinement step. Cost price calculations in the Quick Scan
Design session	Specific instance of a workshop. Joint effort to design the Quick Scan	Relevant to all domains and all steps
Action research	Action and critical reflection in turns. The reflection concerns review of the previous action and planning the next one	Iterative business model design following the actual service design. Relevant for all domains and all steps

Derive Insight into the Viability of a Service Concept

A Quick Scan, i.e., the design of a first draft of a business model, is typically aimed at gaining insight into the viability of a service concept, to decide whether or not to continue developing the service. Typically, this involves some desk research, with a focus on competitive services and price levels, followed by a design session, during which participants together fill in the domains designs of the STOF model. Ideally, such a design session involves four to six participants from different background, e.g. marketing, technology and finance. Ideally, the participants represent potential partners of the service provisioning network. Alternatively, external experts can be invited to fill in certain knowledge gaps. During the session, a made-to-measure workbook is used that explains the service concept and the method, and that describes the questions and activities involved.

Business Model Design for a Research Prototype

Often, research prototypes in the service area focus on technology development, with user needs and requirements as input. When one of the objectives is to transfer this prototype to the market, there is a clear need for a business model design trajectory. Typically, the business case for such a prototype is unclear. One of the ways to start such a trajectory is by involving potential providers or stakeholders and using interviews, workshops and/or desk research as a supporting methodology. Desk research is aimed at finding potential partners or stakeholders, and creating business model alternatives. In this respect, the organizational arrangements and revenue models of related services can be used as inspiration. In interviews and/or workshops, the service concept and business model alternatives are discussed with potential providers and other stakeholders. Another important issue addressed in the interviews is the potential role of the interviewees in the service provisioning network. The interviews result in an overview of the advantages and disadvantages of the alternative business models. Additionally, stakeholders can be asked to assess the CSFs with respect to the business model alternatives.

Examples of how to transfer research results to a business model are provided in Chaps. 9 and 14, where mock-ups are used to evaluate the service domain design. Chapter 16 illustrates an approach to design a complete business model from a technological design, using multiple design sessions with experts, supported by desk research. Chapter 17 provides an illustration of a process where a decision-making support tool was used to assess the viability of business model alternatives with respect to the relevant CSFs.

Develop a Service Based on an Existing (Successful) Service

Service concepts are often based on services that are implemented in different contexts or with different objectives. The business models of such services could provide a valid starting point for service concepts that require a business model design. This means that the business models have to be described according to the STOF model, i.e. including the four domains as well as the CDIs. Additionally, some desk research and/or interviews could be conducted to gain insight into the background and context of the existing services, and to generate selection criteria for CDIs. The resulting business model is analyzed with respect to the context of the new service concept, i.e. the differences are identified. A common design process can then commence, starting with a partially filled Quick Scan. This approach can be combined with action research, as illustrated in Chap. 13, where actual service design and business model design are realized via iterations.

Chapter 6 What's Next? Some Thoughts and a Research Agenda

H. Bouwman, E. Faber, T. Haaker, and R. Feenstra

In this chapter, we conclude part I and provide an outline for further research. We add some thoughts about topics that complement the issues discussed in Part I, e.g. with regard to service bundling and service composition, business processes, and the internal and external validity of our approach. We will also introduce the application chapters that will be presented in Part II of this book.

6.1 Summary and Outlook

In part I, we discussed the relevance of service innovation in generic terms. Service innovation is becoming more and more important in our current service-oriented economy. We discussed trends and drivers that stimulate service innovation and defined our core concepts. Our focus is on the design of (mobile) service(s) (bundles) and their underlying business models. At the moment, there is no common framework for service design, nor is there a common framework for designing business models. In the first part of this book we have tried to develop these frameworks.

Chapter 2 discussed the theoretical foundation for our business model approach, defining the business model concept as well as the core concepts and specifying the relevant concepts in the business models domains. We looked at the interdependencies within as well as between the domains. In Chap. 3, we discussed the core concepts from a design perspective, and discussed the Critical Design Issues (CDIs), and the requirements that have to be balanced for the four specific domains: service, technology, organization and finance. We also introduced the more analytical perspective, directed at the explanation of the viability and feasibility of business models, by looking into Critical Success Factors (CSFs). We want to understand what the effects are of specific choices and what really contributes to the viability and feasibility of a business model. In Chap. 4, we discussed the mobile context, looking more specifically what kind of services are relevant, how technology plays a role both as a driver as well as an enabler of services, and we discussed main trends in the organization domain. In Chap. 5, we introduced the STOF design method in more detail, and illustrated the use of this method in a generic way by discussing a case, that was supportive in developing our approach. In the next part of the book we will discuss some real life cases in which the model as well as the design method were used.

Although our approach has been applied in several cases, as we shall see in the next part, we are aware of some limitations to the STOF model and design method. First of all, we adopt a service provider perspective with a focus on realizing customer value. We are aware of user-centered, and even group or we-centered design approaches. However in this book we don't focus on this aspect, although it is an important aspect of our research, see Steen, Van Eijk, Gunther, Hooreman, and De Koning (2005).

Another limitation has to do with service concepts and creating awareness of the issues that are at stake when trying to understand the implications with regard to technical choices and organizational inter-dependencies. The STOF model helps structure discussion concerning the financial issues, i.e. investments, risks and pricing, and helps in the transition from a service concept towards a service that can be implemented. Nevertheless we are aware that, at a practical level, the implementation of the services requires a more detailed analysis of the processes that are relevant, both from a provider's perspective and from the point of view of mobile services that are intended to support business processes. With regard to consumer services this is relevant, although we really have to understand how the use of mobile services can be embedded into day-to-day practices of users as well as of users that work in organizational processes, and where mobile applications have to support them.

In this book, we touch on (mobile) service bundles in several places, but in most cases a single (mobile) service is developed or analyzed. Elsewhere, we studied mobile service bundles for a consumer market (Bouwman, Carlsson, Molina-Castillo, & Walden, 2007). Based on this research, we may assume that services in a common domain can often be bundled in a meaningful way: specifically entertainment services, mobile (travel) commerce services, mobile Internet services, as well as mobile security and localization services. Consumer surveys provide relatively general patterns with regard to what mobile services consumers prefer. More detailed study of service bundling, for instance based on conjoint analysis, result in more concise and coherent bundles (see, for instance,

Bouwman, Haaker, et al., 2007). Starting from definitions of services bundles (what services should be bundled), the relevant question is how bundles can be designed and enabled using common components that are part of or supporting specific mobile service bundles (the how question). Insight in how to bundle may open further research paths.

This brings us to more practical questions regarding implementation. Due to the holistic nature of our approach, we are aware that tools supporting the implementation of services are needed. This is an important topic for future research. In the next section, we describe three core issues we think may merit closer research. In addition to the practical implications, more academic concerns about the validation of the dynamic business model framework, and the effectiveness and performance of the design method need further attention. The predictive validity and robustness of our models also require further research. These issues will be discussed after a discussion of the more practical implications.

6.2 Practical Implications

Three core issues are at stake when discussing the practical implications of our design framework. The first has to do with the support of the service design with more practical marketing-related tools, the second with how bundles of services can be composed, and the third with the embedding of services in the existing business processes of the organizations involved.

6.2.1 Support Tools

The first question is how to get from high-level service definitions and business models to a more specific, down to earth and practical implementation. Our design method provides generic guidelines and approaches, and includes options to use more elaborate, focused and/or detailed methods and tools. In Chap. 5 we give some indications about which methods and tools can be helpful, but we don't offer a systematic overview of how to deal with specific issues. For example, e3value (Gordijn & Akkermans, 2001) provides an ontological approach to modeling networked value constellations. In some of the cases that will be presented in the next part, the need for more detailed methods and tools was felt. In a next phase of our research into service business model, we will provide additional tooling and practical manuals on how to use our design method.

6.2.2 Service Bundling and Service Composition

Composing service bundles requires careful analysis and selection of services that complement or supplement to each other. Empirical research among users helps to gain insight into which service bundles may be favored by users, but still than the question remains how specific services have to be composed, which specific service components are available and which have to be used. Generic service components are for example security, privacy enhancement, and billing. Such service components can be used to constitute a service when combined with specific functional applications.

In general, decisions regarding the composition of a service are taken within a network of different actors and aim at implementing the technology design. Service Oriented Architectures (SOA's) may provide a starting point for the discovery and execution of relevant services. However, an accepted standardized way to describe the characteristics of services is still missing. Recent developments have extended the SOA paradigm with the use of semantics and ontologies in order to be able to define the service characteristics. The paradigm of the semantic web can be used to extend current web services standards that lack the means to express non-functional attributes of services (reliability, availability, response-time, perceived user experience etcetera). These attributes are key to dynamically selecting the services that best meet user needs (Maxmilien & Sigh, 2004).

To constitute services and service bundles use can be made of a service composition method that supports the process of relating several services and service components. During the process it is decided which service components should be included. Such service components are commonly provided by different suppliers, such as GIS-providers providing location information services, and third parties or mobile telecom operators providing generic services like security, privacy-enhancing applications and billing services. The service composition process consists of three stages. These three phases are part of the exploration phase of the development of a service (bundle) and the supporting business model:

- During the *orientation phase*, an initial technical design is developed where services or service bundles to be developed are described only in general terms, and there is an initial idea about the generic, supplementary and complementary services that may be used. Typically such a design is the result of applying the STOF method.
- The *negotiation phase* is where parties decide – in cooperation – which services and service components are needed, who will deliver which

services and what the specifications of these services will be. This is an essential phase, as the actors should have a common understanding of the end service to be delivered, and the role of constituting services in the final end-user service.

- The moment agreement is reached regarding the services to be used and their specifications, the composition can be realized. Subsequently, the *implementation phase* starts, in which the combined services (the composition) are realized and tested in an experimental setting or on a test market.

Reaching agreement regarding service composition is a crucial step in the process. All parties in the value network should be aware of what can be expected, since the impact of a failure can be substantial Technical problems that can occur are for example related to integration of legacy systems and lack of suitable interfaces. Dependencies between services and service components should be clear, consequences of failure should be made explicit and should be discussed in the value network. Conflicting interests between actors might also be a bottleneck.

Regarding the process of service composition, it is important to specify requirements as a first step (Dym, Agogino, Eris, Frey, &, Leifer, 2005). This includes functional and non-functional requirements. The functional requirements all relate to the purpose of the composition, and to the tasks it has to accomplish. The non-functional requirements relate to the Quality of Service aspects, e.g. the overall reliability and availability. Another issue during the service composition process is to consider alternative services and service components, such that, in case one of the generic services fails, alternative compositions can be used. We feel that further research into service bundling and composition is an important next step in research into service innovations, and business models.

6.2.3 Business Models and Business Processes: Towards a Modeling Approach

Starting from the service bundling and service composition perspective, it will be clear that the processes that support the development, execution, operational use and management of services and service bundles are highly relevant. It is necessary to understand business processes in order to describe their relationships with business models. "A process is a specific ordering of work activities across time and place, with a beginning, an end, and clearly identified inputs and out-puts: a structure for action" (Davenport, 1993, p. 5). Business processes are discussed at great length in existing literature. Various researchers consider them in terms of redesign or innovation (for example Davenport; Hammer, 1990; Melão & Pidd, 2000; O'Neill & Sohal, 1999; Teng & Kettinger, 1995). Many studies are also dedicated to the modeling of business processes (for example Bergholtz, Jayaweera, Johannesson, & Wohed, 2003; Dietz, 1996; Gordijn, Akkermans, & Van Vliet, 2000; Law & Kelton, 1982; Melão & Pidd; *OMG*, 2006; Recker, Indulska, Rosemann, & Green, 2005) and the influence of IT (for example Mooney, Gurbaxani, & Kreamer, 1995; Teng & Kettinger). Business processes are the underlying processes responsible for the deliverance of services. Designing business processes, which have to be in place to enable specific services, requires a cross-functional view; this often means that process design adopts a top-down approach, because designers are the only actors with an organization-wide view. This does not mean, however, that other actors should not be involved in the process. The people performing the actual processes are just as important a source of information, and the same applies to customers. To delivering services in today's networked world, inter-organizational processes have to be taken on board, which is why Davenport opts in favor of a 'networked' view of processes. Designing business process involves the modeling of the business process as they are and as they should be. In the next section, we describe process modeling and the various available techniques.

Process modeling. Process modeling involves the graphical represent-ation of actual processes. The various modeling techniques offer a standardized approach to bridging for the gap between the business process design and process implementation (White, 2004). Many techniques involve the use of ICT tools and are developed in computer sciences. Process modeling, however, is somehow distinct in that the modeled phenomena are to be interpreted by humans rather than machines (Curtis, Kellner, & Over, 1992). Different goals require a different view on modeling and different levels and objects of modeling. According to Curtis et al. and Giaglis (2001), there are four perspectives that are most commonly used:

- The *functional perspective* represents the process elements that are being performed, and the flows of informational entities (e.g., data, artifacts, products) that are relevant to these process elements.
- The *behavioral perspective* indicates when process elements are performed (e.g., sequencing), as well as referring to aspects of how they are performed through feedback loops, iteration, complex decision-making conditions, entry and exit criteria, etc.

- The *organizational perspective* refers to where and by whom (which agents) in the organization process elements are performed, the physical communication mechanisms used for the transfer of entities, and the physical media and lo-cations used to store entities.
- The *informational perspective* represents the informational entities produced or manipulated by a process; these entities include data, artifacts, products (intermediate and end) and objects; this perspective also includes the structure of informational entities.

In recent years, many different modeling techniques have emerged, many with a practical origin and few with a theoretical origin. Examples are Archimate (Lankhorst, 2005), BPMN (business process modeling notation), UML (Unified Modeling Language), simulation modeling and soft system modeling. Many more modeling approaches exist, such as ebXML, Action workflow, Business Action Theory, Resource-Event-Agent framework (Bergholtz et al., 2003). However, from a service and service bundling and business models approach, it is important to understand the general background behind these approaches and the degree to which they are applicable for our purposes. Although business models and business process models have different purposes, they are closely related. Choices that are made in one of the models are bound to influence the other model. There is a lack of methods designed to translate business models into business processes. Further research in this area is required. In future research, we will therefore focus on developing of a tool or architectural approach that combines the more abstract STOF model that support service innovation with more practical and operational business processes, in order to model the value, and information exchanges, as well as the exchange of resources and capabilities at a process level, making it possible to relate service business models to the business processes that support them.

6.3 Academic Implications

From an academic point of view, we have only just begun working on the validation of our concepts and model. Moreover, we also are looking for possibilities to reflect on our business model approach in terms of robustness and predictive validity. In this section we deal with these three issues in greater detail.

6.3.1 Business Models and the Testing of Core Concepts and Causal Models: Internal and External Validity

In Part I, we did describe how concepts and ideas with regard to service innovation and business models evolved from definitions of the core concept toward a development and design method, and subsequently the development of causal models. These causal models are not only related to how we can understand the dynamic relationship between external conditions, phases and the customer and network value of our models, but also to the causal relationship between CDIs, CSFs and the value for the consumers and providers of the services. One of the core problems with regard to these analyses is that the number of cases available to us, as well as the nature of the information that we need in order to be able to do rigorous testing, is limited. Cases are not always as rich as we expect them to be, the information we need is only partially available or accessible, specifically with regard to financial data, and cases are of an unequal magnitude or scale (see also Chap. 7). It is clear that further validation is required, but that practical conditions hinder a rigorous validation. Nevertheless, we have the opportunity to test our concepts in an indirect way, by looking at the research perceptions of the business developers, designers and managers involved in projects regarding innovative mobile services and their business models.

6.3.2 Validation of our Design Method

As we have seen, the STOF method is useful in the early phases of innovation: in exploring and contemplating different ideas and options, i.e. the innovation process is modeled as subsequent phases and iterations, where each phase consists of a divergent (generating ideas) and a convergent (choosing and detailing ideas) part (Buijs, 1984; Buijs & Valkenburg, 1996). In the design sessions of the STOF method, insights are used from creative techniques like brainstorming and boundary examination, insights that facilitate – in a relatively 'free format' – the generation and specification of ideas. The method may be applied whenever people set out to envision and develop business models for innovative services that require cooperation between (a number of) parties. As we have seen, the method consists of a number of steps. In Part II of this book, we will show that it can be used to approach a broad range of topics. Thus far, we have conducted a limited amount of work with regard to the evaluation of our approach. In an explorative research project (Bouwman, Faber, &, Van Der Spek, 2005), we have looked at the effect of using specific parts of our method, the role of the facilitator and the nature of the topics that are being discussed, on the performance of our design method. This research, which is both qualitative and quantitative in nature, was based on six sessions in which the STOF method was used. The study confirmed the usefulness and value of the method. The results provide sufficient reasons to continue our business model research in the direction that we have chosen. We have since conducted a larger number of sessions and used the method in a number of projects that resulted in mock-ups, prototypes, and even some services that have been brought to the market. Having said that, we realize that further research is needed into the performance of our approach. Future work will have to focus on of improving the evaluation and validation tools, which in turn will yield improved versions of the STOF method.

6.3.3 Business Models and Wind Tunneling: Predictive Validity

The STOF method includes a validation of robustness of the business model design, mainly focused on the way the design can adapt to external changes. For further testing the robustness of our business model approach and the business models use can be made of knowledge from futures studies and our experiences in analyzing existing business models on the one hand, and the development of business models on the other. Futures research explores, captures and describes possible alternative futures (Miles, Keenan, & Kaivo-Oja, 2002). It has yielded various approaches to investigating future developments, examples of which are scenario analysis, back-casting, road-mapping, normative forecasting and foresight. Increasingly, different methods are being combined (Bouwman & Van Der Duin, 2003). Scenarios provide insight into the way the future may develop, based on clearly defined assumptions concerning the relationship between relevant developments. Usually, these developments are based on input from other methods of futures research, such as trend analysis. Relevant trends serve as the primary axes along which the alternative scenarios are constructed. A proper scenario study addresses criteria like plausibility (scenarios are not science fiction), consistency (preventing combinations of mutually incompatible trends), completeness (scenarios are more than variations on a theme) and the validity of the underlying assumptions (Van Der Heijden, 1996). Business model evaluation with scenarios is relevant when future market circumstances, technology developments, regulations etcetera, are uncertain, which is often the case for the mobile industry. Scenario methodology is a powerful tool for thinking through the implications of business model choices. Rather than assuming a fixed (often implicitly assumed) future, scenarios offer a range of possible outcomes used not as predictions or forecasts but more as 'wind tunnels' for plans or policies. Using scenarios the uncertainty regarding the future marketplace can be harnessed, and business model choices can be evaluated and possibly adapted. The business models are described in terms of design choices within the business model domains. Future scenarios are defined in terms of the outcomes of scenario indicators. We can assess the impact of the indicator outcomes on the business model design choices. Such analysis provides insight into design choices which can be regarded as robust, i.e. choices that are valid in multiple future scenarios. If certain design choices are not robust, they may be changed to provide more carefully tuned business models that fit certain future scenarios. This provides the flexibility to adapt a business model to changes in the environment.

Furthermore, we feel that the STOF model and method provide important elements in the strategic analysis of scenarios. Evaluation based on scenarios allows us to determine what the relevant concepts are and how they are interrelated, and how they can be used in everyday practice, which in turn helps us improve our method of design. This is one of the ways we test the usability of the method, the relevance of our conceptualization and the robustness of our approach. We have used this approach in a specific case that deal with services of intermediaries in the financial service industry (Bouwman et al., 2005), in which it proved to be very valuable. Chapter 11 illustrates the use of scenarios for business model evaluation for the IPTV business model domain. Generally speaking, it is our impression that scenario analysis helps establish the robustness of a business model.

6.4 Introduction to Part II

In Part II of this book, we shift our focus from the description and conceptualization of our business model framework toward the actual application of the STOF model and method. The cases themselves, and the goal and results of applying our model and/or method, vary considerably. The diversity of applications are proof of the generic applicability of our approach to the business models development in different mobile services domains and sometimes even beyond, i.e., for e-services in general. In all, we have included eleven cases. Some applications have as their primary objective the validation of (parts of) the STOF model. In Chap. 7, the validation of (critical) design issues and success factors for business model viability is addressed from a practitioner's point of view. The question is whether practitioners actually agree with the CDIs and CSFs we have identified and presented in Chap. 3. A large-scale international survey among operators, service and content providers, application developers and experts dealing with mobile services sheds light on this question. In Chap. 8, we look at how the dynamic STOF model works in practice. It shows how the dynamic business model framework discussed in Chap. 2 can help in analyzing the dynamics of business models in a structured approach. As such, it provides a validation of the relevance of the external drivers for change in a business model, i.e. technology, regulation and market, in the phases of the business model life cycle that we identified. Other case studies use the STOF model mainly for analytical purposes, e.g. to identify the most important CDIs or to assess the robustness of business model design choices. In Chap. 9, the service concept and underlying business model for a context-aware, we-centric service for police officers is considered. The we-centric concept and the specific police context point to *Creating Value Elements* and *Network Governance* as the most important CDIs. Mobile payment services, in Chap. 10, have two types of customers (buyers and merchants) and require service providers to find a balance between the interests of end-users, merchants, operators and financial institutions. We use the STOF model to analyze how three different mobile payment providers tried to position themselves in an emerging mobile payment market. Chapter 11 concerns the evaluation of the robustness of mobile service business models, which is a difficult task, albeit a necessary one when it comes to achieving some predictive validity according to the discussion in this chapter. In this chapter scenario analysis is used to assess the robustness of business model for digital television services. Chapter 12 deals with mobile service bundles. Mobile services may sometimes be bundled into meaningful service bundles, e.g. to satisfy related user needs. From a business model perspective, both what should be bundled, and how it should be bundled are relevant questions, as discussed earlier in this chapter. In Chap. 12, these questions are addressed by looking at CDIs that relate directly to the bundling of mobile services. The 'what' question leads to issues like bundle composition, whereas the 'how' question refers to issues like service integration and bundling strategy.

Finally, we have considered cases in which the STOF method is primarily used to design business models for specific service concepts. Often, the design of the business models is followed by some form of analysis, i.e. cross-case comparison (when multiple service concepts are involved) or within-case comparison (when business model alternatives are compared).

In Chap. 13, the STOF method is used to design a business model for the introduction of a mobile remittance service in a developing country.

The design was preceded by an analysis of the business model of a remittance service in another developing country. The analysis of the existing service on the basis of the STOF model proved helpful in identifying and guiding the design of the most relevant CDIs. It provides an example of how existing business models may be analyzed and translated to new contexts. Chapters 14–17 address the important question of how to move from a research mock-up or prototype to actual market exploitation. This is quite problematic in practice as many prototypes never make it past the pilot phase, and financial returns on investments in innovation are often perceived as being too low. A more systematic approach to new service development and underlying business could be beneficial here. Chapters 14–17 provide examples of how the STOF model and method may at least partially provide such a systematic approach. In particular, Chap. 14 shows how the STOF method can be used to assess the business potential of the kind of project mock-ups that are often created in technology-oriented projects for the purpose of demonstration or user testing. However, the mock-ups can also be analyzed from a business perspective. The analysis reveals, for example, which mock-ups should not be pursued, given the apparent lack of business potential. In Chap. 15, a music vending service for a newly industrialized country is examined, in which the STOF model provided the relevant questions with regard to the design of the service; in particular the CDIs were carefully considered. Chapter 15 also provides insight into how the information needed to answer the questions was obtained, e.g., via interviews and user prototype testing. In Chap. 16, we look at a mobile health service, using the STOF method to bridge the gap between a research prototype and an actual service offering. Basically, the approach used is the Quick Scan we discussed in Chap. 5. Several business model design sessions were organized to discuss and identify the necessary design variables. An interesting observation was that the existing prototype is a high-end solution that may not be suitable for commercial exploitation, at least not with regard to the basic service considered most viable in the near future. Chapter 17, finally, provides another example of a healthcare service, albeit not of a mobile nature. In this case, the STOF model was used to design and compare the value networks of alternative business models for a personalized information system providing tailored information regarding services for people suffering from dementia and their informal caregivers. Even though the customer and societal value of such a service may be evident, the design of a viable business model and a business case is problematic. In particular the CSFs were used to assess alternative business models with different focal actors. A decision support tool aided the practical implementation of the evaluation of the CSFs.

Table 6.1 summarizes all application, showing the primary purpose of each case (validation, analysis, design), as well as other case characteristics, i.e. the number of services considered in each case (single, multiple), the degree of domain specificity (generic, specific), the focus of the cases (business model domains, CDIs, CSFs), and finally the target market\b2b, b2c).

Table 6.1. Overview of the applications of Part II and their characteristics

	Chapter	Purpose	Single or multiple services	Domain specificity	Focus	Target market
7	Critical design issues and success factors: practitioner's view	Validation	Multiple	Generic	CDIs	b2c
8	The dynamic business model framework in practice	Validation	Multiple	Generic	Domains and CDIs	b2b and b2c
9	A we-centric service: the PolicePointer	Analysis	Single	Specific	Domains	b2b and b2c
10	Balancing customer and network value in mobile payment services	Analysis	Multiple	Specific	Domains and CDIs	b2c
11	Robustness of IPTV business models	Analysis	Multiple	Specific	Domains	b2c
12	Mobile service bundles	Analysis	Multiple	Generic	CDIs	b2c
13	Mobile remittance services in developing countries	Design	Multiple	Specific	Domains	b2c
14	Assessing the business potential for new mobile services from mock-up evaluation	Design	Multiple	Specific	Domains	b2c
15	Digital music vending machine	Design	Single	Specific	Domains and CDIs	b2c
16	From prototype to exploitation: mobile services for patients with chronic lower back pain	Design	Single	Specific	Domains	b2b & b2c
17	From prototype to exploitation: Organizational arrangements for Personalized Dementia Directory	Design	Multiple	Specific	Domains and CSFs	b2c

Part II starts with the chapters on the validation of the STOF model. The subsequent chapters focus on the analytical uses of the STOF model and method. The final chapters focus on the application of the STOF method in designing business models.

Part II Applications

Chapter 7 A Practitioner View on Generic Design Issues and Success Factors

M. De Reuver and H. Bouwman

What are the generic design issues and success factors for mobile business models that practitioners find the most critical? In Chap. 3, we defined Critical Design Issues (CDIs) as topics that need to be addressed in the service, technology, organization and finance domain of business models. Ultimately, Critical Success Factors (CSFs) predict whether a business model will be viable, i.e. whether it will create sufficient value for customers and capture sufficient value for the providers of the service. Generic design issues and success factors are in contrast to critical design issues and success factors not directly related to a specific to be designed service. This chapter deals with practitioners' views on such generic issues and factors.

The design issues and success factors are derived from a set of exploratory case studies (see Chap. 3). The question is whether they are supported by a broader group of practitioners, dealing with business models that we did not study before. Do the various types of players in the industry agree on which generic design issues and success factors should be addressed? And are they of equal importance throughout the business model life cycle? In this chapter, we address these questions by presenting the results of an international survey among operators, content providers, application developers and experts dealing with mobile services.

We start with a brief discussion of the methodology underlying the survey. Next, we present the results regarding design issues, discussing their relevance in general and in relation to the type of organization and the phases in the business model life cycle. Similarly, we present the results for the success factors.

7.1 Research Methodology

The data were collected through an online questionnaire between September and November 2007. The respondents were presented with a questionnaire containing the list of design issues and success factors from

Chap. 3. They were asked to rate the extent to which they take these issues into account when designing services. To put the questions into context, we asked respondents to focus on their most important service offering. Additional questions were included regarding the type of organization of the respondents.

Finding respondents for such types of surveys is challenging, keeping in mind that there is no database that contains all the relevant practitioners in the mobile services industry. In total, 521 directed invitations were sent out, resulting in 137 business participants and 16 academic experts. The reasons provided for not taking part in the survey were lack of time, lack of expertise to answer the questions and no interest in the study. A specific group of non-respondents consisted of hardware providers and network manufacturers, who commented they did not feel involved in mobile services, but only with technology platforms. Several academics also turned down our invitation, predominantly because they felt they had insufficient expertise to answer the detailed survey questions.

The final sample contained 153 respondents, 78% of whom came from industry, with 22% academic and consultancy experts. Although the survey targeted an international audience, most respondents are from the Netherlands (64). Other regions included in the sample are Scandinavia (18), Germany (9), USA (8), Austria (7), UK (6), Italy (6), France (3), Latin-America (2), Asia (1), Australia (1), South-Africa (1) and other European countries (7). Most industry respondents work at organizations that have been active in the mobile domain since 2000 (65%), while 35% work at organizations that were active even before that date. Similarly, 41% of the respondents were active in the mobile field before 2000, while 59% entered the domain at a later date. Our sample represents a wide variety of 'most important services', including advertising, banking, blogging, communication, e-mail, entertainment, erotic, games, health, Internet, location-based services, news, office, portal, radio, sports information, streaming, surveys, transport information, TV, user-generated content, weather information and workforce management. Of the total number of respondents, 32 adopted the point of view of a (virtual) network operator, 20 that of an application/software provider, 27 that of a consultancy firm, 31 that of a content/service provider, publisher or content aggregator, and only three that of a hardware/equipment manufacturer.

Of the organizations in our sample, 77% collaborate with other organizations to provide a service. They interact on a day-to-day basis with no (27%), one (20%), two (19%), three (15%), four (6%), five (4%) or even more (10%) organizations.

The survey included questions relating to the respondent's core activities, to determine in which phase of the business model life cycle the companies for which they work find themselves (see Chap. 2). Of those that completed these questions, 18 services are in the Technology/R&D phase, 29 in the Implementation/Roll-out phase, and 33 in the Market phase. Strikingly, 74% of the content/service providers are in the Market phase. Apparently, content providers choose as their 'most important service' one that is in its commercial stage rather than one that is still being developed or rolled out. With regard to application developers, 57% are in the Implementation/Roll-out phase. Probably, they choose these types of services as their most important services as implementation and integration are their core activities. In this phase testing first versions of the application and rolling out technology are core issues. As far as operators and consultants are concerned, the number of respondents per phase is balanced.

7.2 Design Issues

This section presents the results of the generic design issues' importance based on 119 completed questionnaires. The respondents were asked to what extent they took the design issues into account, on a 7-point scale ranging from 'not at all' (1) to 'great extent' (7).

The most important design issue in the service domain is *Defining the Target Group* (see Table 7.1). The second most relevant issues are *Trust* and *Branding*. Overall, all design issues appear relevant, as their score is remarkably higher than 4. The standard deviation varies between 1.2 and 1.6 for the service domain issues, which indicates mixed opinions.

In the technology domain, the most important issues are *Quality of Service* and *Accessibility for Customers*. What is interesting here is that *Accessibility for Customers* is viewed as being considerably more important than *Accessibility to Content Providers*. *User Profile Management* is considered less important. Presumably, many of the services that are currently being developed do not yet use profiling techniques. It is also possible that these techniques are not being used because authenticating users is difficult for content providers. The design issue *Security* shows a remarkable pattern: while the average rating is only 4.9, the rating that is given most often is 7, i.e. 'Great extent'. This indicates that opinions among the respondents vary considerably. The average score for technology design issues exactly equals the average score in the service domain, which means that the two domains can be considered equally important.

Table 7.1. Importance of design issues – service and technology domain (*N* = 119)

Service design issue	Average	Standard deviation	Technology design issue	Av	Stdev
Defining the Target Group	5.60	1.20	Quality of Service	5.89	1.08
Trust	5.32	1.31	Accessibility for Customers	5.74	1.14
Branding	5.31	1.28	Scalability	5.19	1.51
Personalization	5.13	1.47	System Integration	5.10	1.41
Unobtrusive Customer Retention	5.06	1.39	Use of Accepted Standards	5.04	1.49
Versioning	4.61	1.61	Security	4.93	1.63
			Accessibility to Content Providers	4.77	1.49
			User Profile Management	4.68	1.66
Average	5.17		Average	5.17	

Generally speaking, design issues with regard to the organization domain are considered to be less important than those in the service and technology domains (illustrated by the lower average rating and higher standard deviation on all items in Table 7.2). Especially *Outsourcing* is considered relatively unimportant, although the high standard deviation indicates mixed opinions among the respondents. More important design issues in this domain are *Customer Involvement* in the service domain and *Partner Selection* in the organization domain.

Table 7.2. Importance of design issues – organization and finance domain (*N* = 118)

Organization design issue	Average	Standard deviation	Finance design issue	Av	Stdev
Customer Involvement	4.95	1.48	Pricing	4.91	1.67
Partner Selection	4.87	1.47	Investment Planning over Time	4.30	1.40
Network Governance	4.68	1.35	Valuing Contributions and Benefits	4.22	1.54
Orchestration of Activities	4.61	1.39	Division of Costs and Revenues	4.21	1.48
Network Openness	4.53	1.55	Division of Investments	3.94	1.35
Outsourcing	3.83	1.67			
Average	4.75		Average	4.32	1.67

Finally, in the finance domain, *Pricing* stands out as the key issue. Topics relating to dividing investments, risks, costs and revenues among partners are considered less important. However, it should be noted that more than a quarter of the respondents indicate that, because they do not work together on a day-to-day basis with other organizations in providing the service, this type of financial issues is less important to them. The average ratings in this domain are lower than in the other domains.

Overall, we find evidence that suggests that all design issues are relevant. With the exception of two of the design issues (i.e. *Outsourcing* and *Division of investments*), all are given a score above 4. This indicates that the respondents do take them into account when they design a service. We find that service and technology-related issues are perceived as being slightly more important than organization issues and far more important than finance issues. Given that service and technology issues are related to creating value for customers rather than capturing value for the providers, organizations appear to focus more on value creation than on value capturing.

Based on the fact that organization and finance-related design issues are given less attention, one might argue that this explains why many mobile services turn out to be less profitable than expected. While organizations seem more comfortable when they are dealing with service and technology issues, the more strategic organization and finance-related issues do not appear to receive sufficient attention. This may indicate that there is a need for greater awareness with regard to organization and finance-related issues, and for the development and diffusion of theories and tools to deal with these issues. On the other hand, the fact, that more than a quarter of the respondents does not work together with other organizations in providing the service, may also explain part of the lower ratings.

7.2.1 Importance of Design Issues Related to Type of Organization

Different players care about different issues. When comparing the ratings of design issues, there are variations among the four most prominent organization types in our sample, i.e. operators, content providers, application providers and consultants. The variations are presented in Fig. 7.1, on a 7-point scale for importance, ranging from 'not at all' (1) to 'great extent' (7).

In the service domain, it appears that *Branding* and dealing with different versions of the service *(Versioning)* are considered less relevant by consultants. On the other hand, *Personalization* is considered more important by application providers. This may be because personalization is especially driven by advances in software platforms that are dealt with by application providers.

Although in the technology domain there are not many differences, it is interesting that content providers do not take *System Integration* and *Security* into account to the same extent that the other organizations do. This could be because these issues are typically solved within applications or within an operator's network, which means it is less of an issue as far as content providers are concerned. Another difference is that operators do not take the use of accepted standards into account as much as the other organizations do. This may be because operators tend to dictate billing, authentication and other standards to which content providers and application developers are expected to adhere.

In the organization domain there are few differences of opinion. One remark can be made about *outsourcing*, which appears to be much more important to operators than it is to application providers. This may have to do with the fact that application providers are the ones who deal with the outsourced activities relating to software and platform development, and as a result they are less likely to perceive outsourcing as something they deal with themselves. On average, operators find organization design issues slightly more relevant than the other players, but this difference is statistically insignificant.

As far as financial issues are concerned, *Pricing* is more of a core issue to operators than it is to the other players. Operators often assume the role of billing and collections provider, and as such they determine the prices that service providers are able to charge. Like in the organization domain, finance-related issues appear to be more important to operators than to the other players.

The five most important design issues relative to the organizational type is presented below. It is interesting to see that consultants find technology-related issues the most important. As far as the other players are concerned, a mix of service and technology related design issues are considered the most critical.

	Operator	Content/service provider	Application/software provider	Consultancy
1	Branding	Accessibility for Consumers	Personalization	Quality of Service
2	Quality of Service	Quality of Service	Clearly Defined Target Group	Accessibility for Consumers
3	Clearly Defined Target Group	Branding	Accessibility for Consumers	System Integration
4	Pricing	Trust	Quality of Service	Security
5	Accessibility for Consumers	Clearly Defined Target Group	Scalability	Clearly Defined Target Group

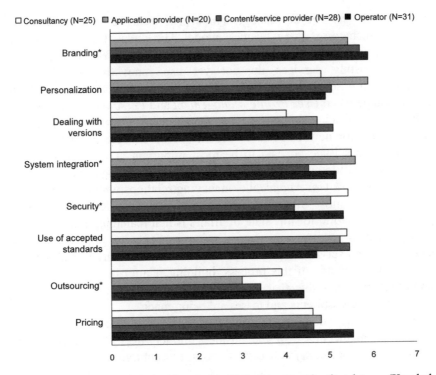

Fig. 7.1. Importance of design issues in relation to organizational type (Kruskal–Wallis test, $* p < 0.05$)

7.2.2 Design Issues and Business Model Life Cycle

Some of the decisions concerning a new business model have to be made on day one, while others can be postponed. In this section, we discuss the relevance of the design issues within each of the three phases of the dynamic STOF model, i.e. Technology/R&D, Implementation/Roll-out, and Market. The most interesting differences are presented in Fig. 7.2.

In the service domain, we see that *Branding* is relevant in the first stage of a new business model, and becomes even more important when the service is rolled out and commercialized. Because branding a service is especially important when it is offered on a commercial basis, this makes sense. On the other hand, *Personalization* appears to become less relevant in the roll-out phase, possibly because services that use advanced forms of personalization are probably still in their pilot phase and are not yet widely adopted.

The attention to technology design issues does not increase throughout the product life cycle, opposite to what one might expect. They are addressed at the start of developing a business model, and remain relevant over time, with exception of *User Profile Management*, which seems to be most important in the initial stages of a new business model. The explanation here is similar to the one we suggested with regard to *Personalization* in the service domain: existing services do not yet use innovative profiling and personalization technology. This explanation is supported by a strong correlation between *User Profile Management* and *Personalization*.

In the organization domain, we see that *Network Openness*, i.e. offering opportunities for new partners to become involved becomes less relevant over time. This issue appears to be most important at the start of a new business model life cycle, when there is a quest to find partners in the value network. *Network Governance*, i.e. managing relations with partners, on the other hand, is most important in the Implementation/Roll-out phase. A possible explanation for this state of affairs is that managing relationships is most crucial when moving from service concepts to implementing the technology and service in practice, because there are many things that may happen in the course of this process, building relationships becomes more important. It is possible that the stakes go up when a service is actually launched and investments and losses are being made.

In the finance domain, the relevance of issues hardly seems to be dependent on the phase of the business model life cycle. Only *Valuing Contributions and Benefits* becomes less important over time.

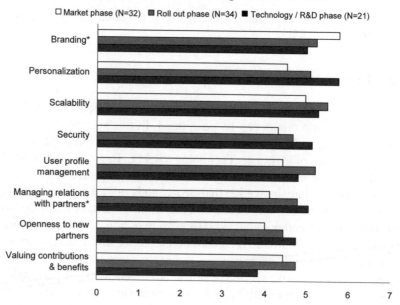

Fig. 7.2. Importance of design issues in relation to life cycle phase (Kruskal–Wallis test, * $p < 0.05$)

A top-5 of most important design issues for each of the phases in the business model life cycle is presented below. Generally speaking, most of the design issues are equally relevant in all the phases. *Quality of Service*, for example, appears the most critical issue in all phases.

	Technology/ R&D phase	Implementation/ Rollout phase	Market/ Commercial phase
1	Quality of Service	Quality of Service	Quality of Service
2	Clearly Defined Target Group	Accessibility for Consumers	Branding
3	Personalization	Scalability	Accessibility for Consumers
4	Accessibility for Consumers	System Integration	Customer retention
5	Generating trust among customers	Clearly Defined Target Group	Clearly Defined Target Group

7.2.3 Success Factors

Respondents were asked to what extent they included success factors as objectives for their service offering. For each of the factors, a number of items were assessed, which we developed ourselves, except for *Unobtrusive Customer Retention* that was derived from Appiah-Adu, Fyall, and Singh (2001). After an exploratory factor analysis, each factor was assigned the average score for the corresponding items, ranging from 'not at all' (1) to 'great extent' (7). Table 7.3 provides an overview of how our respondents evaluate the success factors. We draw a distinction between success factors leading to value creation for customers, and those leading to value capturing for the network of service providers, i.e. network value. With regard to network value, we found that items relating to *Acceptable Profitability* and *Acceptable Risks* are strongly related. Respondents that stress profitability related objectives are likely to find risk related objectives important as well. Consequently, we combine these success factors in the analysis.

Generally speaking, all success factors have a score exceeding 4, implying that they are considered relevant. With regard to customer value, the major success factor is having a *Compelling Value Proposition*. Ratings were very high on the items relating to this success factor, such as a clear definition of added value and clearly communicating the value of the service. *Clearly Defined Target Group* is also important. The ratings for *Acceptable Quality of Service* and *Unobtrusive Customer Retention* are lower, but still above average.

Generally speaking, success factors relating to capturing value for the providers are rated slightly lower than those relating to value creation for customers. This seems to be in line with our findings with regard to the design issues, where service and technology-related issues score higher than organization and finance-related issues. What is most important is having an *Acceptable Division of Roles*, followed by *Acceptable Profitability* and *Acceptable Risks*.

Table 7.3. Importance of success factors

	Success factor	Average rating	Standard deviation	N
Customer Value	Compelling Value Proposition	5.74	0.83	122
	Clearly Defined Target Group	5.24	1.20	123
	Acceptable Quality of Service	4.88	1.27	120
	Unobtrusive Customer Retention	4.48	1.22	119
Network Value	Acceptable Division of Roles	5.05	1.04	119
	Acceptable Profitability and Risks	4.89	1.11	119

To summarize, the top-3 success factors for mobile business models are:

1. Compelling value proposition
2. Clearly Defined Target Group
3. Acceptable division of roles

7.2.4 Importance of Success Factors Related to Type of Organization

One might expect the various players to pursue different goals with their mobile services and business models. However, when we compare the ratings for the four types of organizations in the sample (i.e. operators, content providers, application providers and consultants), then we don't find statistically significant differences. There is an indication, however, that operators consider *Acceptable Quality of Service* more critical than content providers and application providers. This may be due to the fact that operators are more responsible for the actual quality of service, and because they are more directly involved with issues relating to the quality of the service system, such as security, authentication and end user privacy. The top-3 success factors per organization type are:

	Operator	Content/service provider	Application/software provider	Consultancy
1	Compelling Value Proposition	Compelling Value Proposition	Compelling Value Proposition	Compelling Value Proposition
2	Acceptable Quality of Service	Clearly Defined Target Group	Clearly Defined Target Group	Acceptable Division of Roles
3	Clearly Defined Target Group	Acceptable Division of Roles	Acceptable Profitability and Risks	Acceptable Quality of Service

7.2.5 Success Factors and Business Model Life Cycle

Although one might argue that the objectives of service offerings shift over time, we find only slight variations with regard to the importance of success factors across the phases in the business model life cycle, see Fig. 7.3. The only significant difference is *Clearly Defined Target Group*. While this is the second most important factor in the Technology/R&D phase, it becomes one of the least important factors in the Market phase. Apparently, it is the most relevant in the early stages, becoming less important once the target group has been established. It is also possible that finding niche markets is more important for newly developed services, compared to mature services that expand over time from niche to mass market services, making targeting less of an issue. There are slight variations across the phases as far as the other success factors are concerned.

The top-3 success factors per phase is given below. This list illustrates the finding that the importance of success factors hardly depends on the phase in the business model life cycle.

	Technology/ R&D phase	Implementation/ Roll-out phase	Market/ Commercial phase
1	Compelling Value Proposition	Compelling Value Proposition	Compelling Value Proposition
2	Clearly Defined Target Group	Clearly Defined Target Group	Acceptable Profitability and Risks
3	Acceptable Division of Roles	Acceptable Division of Roles	Dominant Position by Taking Risks

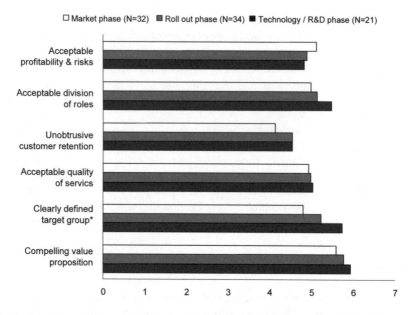

□ Market phase (N=32)   ■ Roll out phase (N=34)   ■ Technology / R&D phase (N=21)

Fig. 7.3. Importance of success factors related to life cycle phase (Kruskal–Wallis-test, * $p < 0.05$)

7.3 Conclusions

Generally speaking, we find that all generic design issues and success factors are considered important by the respondents, although some are considered more important than others. Overall, service and technology-related design issues are perceived as being more important than organization and finance-related issues. As far as several design issues are concerned, there are differences in opinion between the various actors in the value network (i.e., *Branding, Versioning, System Integration, Security,* and *Outsourcing*). We also find that some of the design issues become more relevant as a service moves through the various phases of the business model life cycle (i.e. *Branding*), while others become less important over time (i.e. *Personalization*). According to the survey results, the top-10 most important design issues are: *Quality of Service, Accessibility for Consumers, Clearly Defined Target Group, Trust, Branding, Scalability, Personalization, System Integration, Unobtrusive Customer Retention,* and Use of Accepted Standards.

Success factors relating to value creation for customers are slightly more important than those related to capturing value for service providers. There is little difference in opinion between the types of organizations in the value network, although *Acceptable Quality of Service* is more important to operators than it is other players. The importance of the success factors does not appear to change significantly during the business model life cycle, with the exception of *Clearly Defined Target Group*, which becomes less important over time. The top-3 critical success factors are *Compelling value proposition, Clearly Defined Target Group,* and *Acceptable division of roles.*

Chapter 8 The Dynamic STOF Model in Practice

M. De Reuver and H. Bouwman

What triggers companies to revise their original design choices in the business model? And what can be learnt from that to make new business models more robust towards a changing environment? As business models are dynamic rather than static, it is important to understand how they evolve over time. Such understanding helps to explain why certain companies are better able to adapt than others, and to make new business models more robust. In this chapter, we show how the dynamic STOF model from Chap. 2 can help in such analyses.

Often, business models are studied through rich and extensive narratives, making it hard to analyze their core elements while taking into account the dimension of time. A structured, repeatable approach would help to understand how a specific business model has evolved over time. Similarly, when comparing the dynamics within different business models, such an approach is beneficial in order to find generic patterns in their evolutionary cycles.

In Chap. 2, we introduced the STOF model that links external technology, market and regulation forces in the environment to internal business model components of service, technology, organization, and finance. Moreover, the model divides the life cycle of business models into three subsequent phases, i.e. Technology/R&D, Implementation/Rollout, and Commercial/Market. In this chapter, we show how the STOF model can be applied in analyzing the dynamics within a single business model (within-case analysis) and in comparing the dynamics of different business models (cross-case analysis). We show both qualitative and quantitative approaches to do so.

We illustrate the use of the STOF model for three specific purposes. First, the use of the model in studying the dynamics of a single business model in-depth is shown. This is helpful from research perspective as the causalities underlying changes can be understood, but also from practitioner point of view to gain more understanding of how their company is shaped by its environment. Second, we illustrate the application of the model in a qualitative analysis of a small set of business models. In this study, both within-case analysis is done to understand the dynamics of each individual business model, and cross-case analysis is applied to compare the business models while taking into account the time dimension. Third, it is shown how the STOF model can be used in quantitative studies on larger sets of business models, in order to perform statistical analysis on the patterns found in cases. The latter approach is applicable for example for testing a theory on what external forces drive the changes in certain design choices.

The approaches in this chapter can be used for academic as well as practitioner purposes. From academic perspective, it can help to structure the findings of case studies on business models while taking into account the time dimension. Once it is recognized that business models are not static but change over time, such time dimension cannot be ignored but should be taken into account in the analytical framework used. In the studies described in this chapter, the so-called case survey method is used as proposed by Larsson (1993) and Yin and Heald (1975). In this approach, existing, rich descriptions of business models are analyzed in a structured protocol. Then, qualitative and quantitative analysis is performed on these findings. However, the approach and the framework are also usable when first-hand case study material is being analyzed.

For practitioners, the approaches are relevant in order to understand how their business models have evolved over time, and how they can make them more robust. It also allows them to compare their business models with other business models, while taking into account situational differences in terms of external drivers and the maturity of the business model.

The studies described in this chapter have been used to validate the dynamic STOF model, and have been part of a three-stage research strategy, starting from one illustrative case to the qualitative and quantitative study of multiple cases (Bouwman & MacInnes, 2006; Bouwman, MacInnes, & De Reuver, 2006; De Reuver, Bouwman, & MacInnes, 2007; De Reuver, Haaker, & Bouwman, 2007).

8.1 Study 1 Qualitative, Single-Case Analysis

Often, researchers and practitioners want to understand how and why a specific business model has evolved over time. What triggers the companies to revise their original design choices in the business model? And what can be learnt from that to make new business models more robust towards a changing environment? We illustrate here how the dynamic STOF model can help in answering these types of questions for a specific business model through an in-depth, qualitative analysis, making use of a structured coding protocol.

The business model studied here is the well-known case of i-mode from NTT DoCoMo which is a mobile Internet portal service offered in the Japanese market. In studying this case, information on external drivers and internal business model components was structured according to the model while taking into account the various phases. Then, it was analyzed what external drivers were most important for this case, and how they triggered changes in the business model design choices.

8.1.1 Illustrative Results

An overview of the structured findings is given in Table 8.1. In the first phase, an important driver was found to be regulation, i.e. liberalization of the telecommunication market that ended the government-owned monopoly of incumbent operator NTT, as well as the decision in 1994 by the Ministry of Post and Telecommunications to allow users to own handsets instead of leasing them. This minor change in regulation opened market expansion as well as technological opportunities. In addition, there were technological developments related to security, reliability, and emergency location. This offered NTT DoCoMo the opportunity to experiment, first by using a dedicated private network and subsequently with a less ambitious exploratory project that took advantage of packet based transmissions to deliver data and allowed users to have an 'always on' connection. In this early stage the company was looking for and experimenting with alternatives that could potentially succeed in the market.

The liberalization of the telecommunications sector allowed companies to expand their markets and many carriers all around the world were looking for business opportunities. This was a time when wireless adoption had accelerated and appeared to have endless possibilities. In Japan specifically the conditions were optimal for a technological breakthrough in mobile connectivity. As such, there is a link between the market and technology environment and DoCoMo's business model.

NTT DoCoMo had identified a promising market segment. It focused its attention on young urban adults with high disposable incomes and a propensity to try new gadgets. In addition there were individuals that spent much time commuting and given their limited living spaces spent much of their time outside in the streets, in restaurants and other public areas. Another factor that contributed to the success of the mobile Internet in Japan was the high price of broadband access at home. A wireless Internet connection was thus a significantly cheaper alternative. This indicates that there is a link between the market developments and the service component of the business model, more specifically the decision on targeting.

Table 8.1. Findings of DoCoMo i-mode case in phase I, structured according to the dynamic STOF model

Technology/R&D phase (1997–1998)		
Driver	Technology	– Development of mobile data services
	Market	– *Competition in wireless voice market*: market share declining from 80 to 60%
		– Young urban adults have high disposable incomes, propensity to try new gadgets, and they commute
		– High price of broadband Internet
	Regulation	– Deregulation of voice market
Compo- nent	Service	– *Target group*: young females
	Technology	– CPE small device, big screen
	Organization	– Task force for data communication opportunities (Keiichi Enoki: business experience; Takeshi Natsunoe: Internet start up; Mari Matsunaga: publishing)
		– *Content Providers*: acquire content, generate internally
		– DoCoMo focus on network
		– Creation of Gateway Business Department as gateway, market maker
	Finance	– Content reselling to customer on basis of subscription

Implementation/Roll-out phase (February 1998–November 1998)		
Driver	Technology	
	Market	– *Initially no public attention*: focus on aggressive marketing campaign
		– 6 months after introduction 2 million subscribers
	Regulation	
Compo- nent	Service	– *Marketing*: role model(s)
		– *Branding*: alliance with Sumitomoto Bank
		– *Broadening services*: critical mass of services
		– Official i-mode sites by 800 content providers, and 40,000 voluntary sites, i-mode compatible in 2001
	Technology	– Virtual private portal for consumer access always on
		– cHTML for access of content, in combination with HTTP and SSL
		– Java J2ME as script language
		– Choice for JAVA middleware above MPEG-4; BREW platform
		– *Wireless phone system*: established payment mechanism

	Organization	– *Walled garden approach*: DoCoMo center of gravity – Balance value web, incorporating component manufacturers, handset manufacturers (specifications defined by DoCoMo), equipment manufacturers (W-CDMA technology), Service Providers (JAVA-based), Content Aggregators; Enterprise Solutions Providers and Content Providers
	Finance	– Cross subsidization by DoCoMo of handset (between ¥ 20,000 and ¥30,00) – Fixed fee (¥ 300), charge for data packages (¥ 0.3 per packet). Lower than price for ISP (¥ 2,000) – *Official i-mode sites*: fee from subscribers via telephone bill – *Voluntary sites*: collect fee themselves – 9% of official content providers revenues for DoCoMo – Advertisement revenues – 30% of i-mode sites (premium) charge subscribers – Risk reduction by DoCoMo if subscriber defaults on payment

Market phase (2001 – present)		
Driver	Technology Market Regulation	– Development of mobile data services – 30 million users in 2001; 38 million in 2003 – FOMA services 330,000 (April 2003) – *i-mode users*: higher retention rates – Easy switch between content providers enabled by DoCoMo
Compo- nent	Service	– *Broad acceptance*: market share of 56% in 2001 – Voice data board accepted
	Technology	– Data-rate 28.8 kbps – Shift to 3G (from 9.6 kbps to 64 kbps, moving up to 384 kbps), Vertical applications i.e. M2M applications, LBS, Camera, Streaming video. Freedom to Multimedia Access concept (W-CDMA) – Network problems; congestion due to traffic demand, 3G as alternative solution. Trials show problems with handset and poor connection rates
	Organization	– Relationships with content providers redefined: not only the walled garden approach, but also a more open format for content providers was defined
	Finance	

As indicated by the DoCoMo case, the STOF model can be used to structure the information on business model dynamics. Structuring the findings across phases allows seeing how the business model has developed over time, and what changes have been made in it. While doing so, it provides more insight in the dynamics of a business model than using static descriptions of business model components that tend to focus on the mature stages of successful business models solely. In addition, as external forces are included in the analysis, interpretations can be made on which external drivers are likely to have caused which internal changes in the business model components. As such, the approach provides a far more structured approach in studying the dynamics of a single business model than plain narratives.

8.2 Study 2 Qualitative, Multiple Case Analysis

While the in-depth study of a single business model can provide valuable insights, broader insights can be gained by contrasting different business models. In a second study, six business models were studied, including i-mode. By contrasting the findings, the importance of the drivers across cases was studied. In addition, links were studied between external drivers and the organizational arrangements in the business models. As such, the study entailed both cross-case and within-case analysis.

To ensure maximum insight from a limited number of business models, one can select those that differ regarding important dimensions. In this study, the disruptiveness of the innovation and the type of industry were used as differentiators. Table 8.2 provides an overview of the cases studied.

Table 8.2. Cases for qualitative analysis

		Industry type	
		Mobile business	Other sectors
Innovation type	Disruptive	NTT DoCoMO's i-mode	MySQL Skype
	Sustaining	Telmore	C-Commerce Kodak

The cases were coded by three observers. After training them and comparing the results of a first case in order to see if any problems occurred, they analyzed the cases independently. While coding the cases, they used a coding protocol similar to that in the first study. However, this protocol also provided checklists of the types of aspects they could find in the cases regarding the drivers, components and performance issues.

One of the cases, i.e. Kodak was eliminated because multiple services and products were introduced, while none succeeded to enter the second phase.

8.2.1 Illustrative Results

The findings of the qualitative schemes were summarized and structured according to our concepts, see Table 8.3 for a fragment of the resulting table. For example, it was found that in the Telmore case important drivers in the first phase were mainly related to market (i.e. a highly competitive market; the vertical integration of mobile network operators) and regulation (i.e. the regulator forced the incumbent operator to open its market to virtual network operators; handset subsidization). These drivers led to the choices in the Telmore business model which focused on being a mobile virtual network operator with far lower costs than incumbent operators in order to survive in the highly competitive market.

Then, these qualitative findings were coded into ordinal scores. The importance of the drivers was coded and it was assigned whether the business model component had changed compared to the previous phase. It appeared that at least in these cases, the assumptions of the model regarding the importance of drivers were not completely correct, as all drivers were important in the first phase; in the second phases there are almost no external drivers at all; and in the third phase, the few drivers that were found are of moderate importance.

The STOF model was also used to analyze the organizational changes in more detail. And then, an interpretation was made which types of drivers would have led to these changes. For example, in the i-mode case it was found that the walled garden strategy of the company was loosened in the third phase and that DoCoMo had to allow other portals on their network, basically as a result of regulatory obligations. Another interorganizational governance related change was the partnering of i-mode with handset makers in the first phase, which was probably due to deregulation that allowed users to own their handsets instead of leasing them. And in the third phase, it was found that i-mode partnered with international companies, probably as a result of market saturation and the rise of two strong competitors on the home market. Similar analyses were made for the other cases, leading to in-depth understanding of how organizational arrangements were forced to change by external drivers.

Table 8.3. Findings in qualitative study of five cases (fragment)

	MySQL – phase I	Telmore – phase I	i-mode – phase I
Drivers			
Market – competition	Growing open source phenomenon Only large vendors offer Database management software	Highly competitive environment in Denmark Mobile operators are vertically integrated (high costs)	Fierce competition Decreasing market share Preparation of DoCoMo for the future
Market – customer			Increasing Internet adoption PC adoption lagging behind Rapid cellular phone adoption
Technology	Internet used as a business tool Internet expansion leading to heterogeneous components and hindering consolidation		PHS phones with 64 kbps transmission capability
Regulation	Evolving diversity license agreements for open source software, enabling various strategies	Danish regulator mandating TDC (Danish incumbent) to sell wholesale network access Handset subsidization	Market being opened for foreign players leading to competition Liberalization of mobile market Deregulation allowing customers to own handsets Limited spectrum allocated leading to innovation
Business model concepts			
Service	*New value proposition*: Simple and easy to install DBMS *New market segment*: unsophisticated users New targeted user experience	*New value proposition*: Basic voice and SMS service offered through simple Internet based model *New differentiation*: low pricing	New value proposition New market segment New market strategy New pricing strategy
Technology	*Application*: Incremental, LAMP development (Linux, Apache, MySQL, PHP) as building blocks	*Network*: TDC infrastructure *Application*: SMS/Internet *Device*: SIM	*Network*: Old network *Application*: New cHTML protocol, GIF, Midi, email, emoticons *Device*: New handsets

The application of the STOF model in the qualitative analysis indicates its usefulness when analyzing multiple cases. Not only is it shown that the model can be used to structure the findings in other cases than the i-mode case, it also shows that both within-case and cross-case analyses can be facilitated by coding findings according to the model. It is both possible to compare the importance of drivers or business model components throughout phases, as well as to interpret how components have been impacted by drivers.

When analyzing qualitative case study findings, richness of material is always an issue, whether the material has been collected first-hand or has been sourced elsewhere. The approach shown here of summarizing qualitative statements in the material with the structured framework, and then analyzing them on predefined metrics, proves useful in making sense of large amounts of rich descriptions. Therefore, qualitative studies on business model dynamics with other research questions can benefit from adopting the approach presented.

8.3 Study 3: Quantitative Analysis

In contrast to the two qualitative approaches described so far, the STOF model can also be used to make a quantitative analysis of a large number of cases. This is particularly useful when a theory on the relations between external drivers and business model design choices already exists, and one is interested in testing the robustness of this theory across a large set of cases.

In the study described here, the question answered was what external drivers would be most important relatively to the three phases in the model in a large set of e-business models. To do so, 45 business models were analyzed using the STOF model, ultimately aiming to establish statistical generalizable cross-case comparisons. First, a very detailed case study protocol was developed, based on the findings of the two previously described studies. This protocol detailed variables that could be coded for external drivers, business model components, and business model performance.

Regarding external drivers, the protocol asked coders to indicate whether a specific driver had been of any importance for the business model in the phase. For example, technology driver variables included general technology trends, i.e. digitization, processing power, miniaturization, mobile technology, technical integration, positioning technology, intelligent systems, interoperability, security, and natural interfaces (Bouwman, Van den Hooff, Van de Wijngaert, & Van Dijk, 2005), as well as Internet technology, standardization bodies, incremental nature of technology, and degree of technical sophistication. For the business model components, the protocol specified specific choices on aspects of the component wherever possible. For a significant part of the component variables no categories could be derived, so free-format coding was allowed followed by content analysis on the coding descriptions.

Analyzing the contents of cases by coders involves certain subjectivity, which was accounted for by having two independent observers. After coding all cases, intercoder reliability statistics where computed using the percent agreement and Cohen's Kappa. Especially for some of the internal business model component variables regarding the organizational architecture, intercoder reliability was too low to allow further analysis. Presumably, this was partially due to loose operationalization and lack of richness in the case descriptions.

8.3.1 Illustrative Results

The research question addressed in this study was which external drivers would be most important in which phases, thus validating the propositions in the STOF model. To answer this question, aggregate measures of the driver variables were computed. These were a ratio-scale measure that simply counted the number of lower-level driver variables of the relevant type that were deemed relevant by the coder, and a binary-scale measure that indicated whether or not any of the lower-level driver variables were ticked in the coding of the case.

In regression analysis, the aggregate measures for the driver variables were used as dependent variable, and the phase of the business model as an independent variable. In doing so, both linear regression (using the ratio-scale measures) and logistic regression (using the binary-scale measures) could be conducted.

An example of resulting output is given in Tables 8.4 and 8.5, which relates the importance of technology drivers to the phase of the business model. These results indicate that technology drivers were far more often found in the first phase of cases, and less in the second and third phase.

Table 8.4. Linear regression for ratio-scale technology driver

	B	SE B	B
Constant	1.222	0.183	
Phase II (dummy)	−0.801	0.255	−0.37**
Phase III (dummy)	−1.005	0.293	−0.38***

Note: $F = 7.4795$; $df = 94, 2$; $p \leq 0.001$. $R^2 = 0.14$; $**p < 0.01$; $***p < 0.001$

Table 8.5. Logistic regression for binary-scale technology driver

		95% CI for exp b		
	B (SE)	Lower	exp b	Upper
Constant	0.223 (0.335)		1.25	
Phase II (dummy)	−1.39** (0.508)	0.092	0.248	0.672
Phase III (dummy)	−2.575** (0.812)	0.015	0.076	0.375

Note: Hosmer and Lemeshow p-value = 1; $R^2 = 0.16$ (Cox and Snell), 0.22 (Nagelkerke). $**p < 0.01$

To account for case study findings that might be biased to background variables, similar regression analysis was carried out splitting the data set in different types of cases, e.g. differentiating according to industry sector or size of the company developing the service. Illustrative results of such an analysis are shown in Tables 8.6 and 8.7. The results indicated that the importance of technology drivers only depends on the phase for startup companies, and that this relation is not found for business models of established companies.

Table 8.6. Logistic regression for binary-scale technology driver, startups

		95% CI for exp b		
	B (SE)	Lower	exp b	Upper
Constant	0.095 (0.437)		1.100	
Phase II (dummy)	−1.992** (0.758)	0.136	0.031	0.602
Phase III (dummy)	−21.298 (10,377.78)	0.000	0.000	0.000

Note: Hosmer and Lemeshow p-value = 1; $R^2 = 0.26$ (Cox and Snell), 0.39 (Nagelkerke). $**p < 0.01$

Table 8.7. Logistic regression for binary-scale technology driver, established companies

	B (SE)	95% CI for exp *b*		
		Lower	exp *b*	Upper
Constant	0.405 (0.527)		1.500	
Phase II (dummy)	−0.811 (0.745)	0.103	0.444	1.915
Phase III (dummy)	−1.504 (0.972)	0.033	0.222	1.493

Note: Hosmer and Lemeshow *p*-value = 1; $R^2 = 0$ .073 (Cox and Snell), 0.098 (Nagelkerke). $p < 0.05$

The study discussed here illustrates how the STOF model may be used in analyzing large sets of case descriptions in a quantitative approach. Research questions answered in this approach can be various, for example on the impact of technology changes on the likelihood of forming alliances in the organizational component of the business model, or on even more complex questions linking drivers, components and performance metrics. By including background variables, one can contrast analyses of different types of cases, thus allowing more specific cross-case analyses.

An advantage of the quantitative approach is of course that larger numbers of cases can be studied, allowing for statistical generalization. However, it is still advisable to approach the research question first in a pilot study, using a smaller set of cases in a qualitative approach, and then use the lessons learned to develop a more stringent quantitative protocol. It is highly important to have rich information in the case description and to operationalize the core concepts to be measured in a strict approach. Failing on these criteria may lead to inconsistencies in the coding of independent coders, and hence unusable results.

8.4 Conclusions

This chapter showed how the STOF model from Chap. 2 can help in analyzing the dynamics of business models in a structured approach. By structuring information about external drivers and internal business model components along the phases of a business model life cycle, the STOF model helps to understand the dynamics within a specific business model. We also showed how the model can be used to compare a limited set of business models in a qualitative approach. Finally, we discussed a quantitative study of a large set of cases, in order to do statistical analysis of the patterns in dynamics of a larger set of business models. In general, we advise combining qualitative and quantitative approaches in these studies.

Chapter 9 A We-Centric Service: The PolicePointer

M. De Reuver and M. Steen

In this chapter, we discuss the design of a service concept and its underlying business model for a context-aware, we-centric service for Dutch police officers. We-centric services are meant to support people in their communication and collaboration in dynamic groups that may change or emerge over time. Typically, these kinds of services add value by locating colleagues, finding out who has relevant information on the user's current situation and/or discovering which group members are available for direct communication. Designing business models for we-centric services is a novel area. As end-users can both consume and provide value for the service offering, unique business model issues emerge especially in the service domain.

In this case, the we-centric service called PolicePointer aims at facilitating the exchange of communication and knowledge between community police officers and emergency police officers. Service innovation within the Dutch police organization is a challenging affair. According to one stakeholder, it typically takes as much as 8 years to develop and implement a new service, largely due to the organizational complexity of the police force. Historically, the Dutch police was highly decentralized. Before 1993, each municipality had its own police force. The 200 forces were merged into 25 regional forces and one special, national force. The 26 'police regions' operate relatively autonomously, without direct guidance from the national government. Over the years, each police region developed its own ICT systems. This led to problems concerning interoperability, as it was often difficult to exchange information between police regions. Because this situation was considered undesirable, in the late 1990s, a central ICT service organization was established, which now owns and manages the existing ICT systems, controls the access to police databases, and provides the mobile communication network C2000.

We begin with a brief description of the process involved in designing the service concept and business model of the PolicePointer. Next, we apply the STOF model to the service concept and discuss the critical design issues in the business model underlying the PolicePointer.

9.1 Designing The Service and Business Model

The PolicePointer was developed in collaboration with the central ICT service organization, in a partly government-funded research and development project. Over a total period of 3 years, the service concept was developed, refined, prototyped and tested, in a user-centered design approach, i.e. the project team members sought interaction with future and potential end-users during the innovation project. As a result, many of the design choices are based on interaction with end-users.

Various steps were taken in the service concept design process, spanning from 2004 to 2006. First of all, a workshop was organized that involved police officers, from which lessons were drawn regarding area-bound work and existing communication, information and cooperation processes in which police officers are involved. Next, problems and opportunities were identified that may be solved through innovative mobile services. From the four opportunities which were articulated in this workshop, one was selected that was deemed to fit the we-centric concept best: the need for community police officers to communicate with other police officers, firemen or ambulance personnel, and with network partners, such as the municipality, social welfare workers or school headmasters. This meant that the design choice that was made at this stage was that the service should support community police officers in communicating and cooperating with network partners, i.e. others outside the police force.

Secondly, a so-called rapid ethnographic study (Millen, 2000) was conducted to observe and understand the daily work of police officers through their personal experience. During these observations, special attention was paid to how police officers communicate and cooperate with others, in different contexts and for different tasks. The observations were written down, and relevant and frequent events were then summarized into personas and storylines, describing a typical working day in the life of typical end-users. In a workshop, these observations were validated by community police officers. Based on the observations and the workshop, the scope of the service concept was narrowed by focusing on supporting community police officers to share their knowledge with emergency police officers specifically.

Thirdly, the resulting service concept was evaluated in two additional workshops involving police officers. In the first workshop, the objective was to validate the problem and the proposed solution as being relevant and valuable to police officers. The design choice that was made after this workshop was to stimulate emergency police officers to access and use the knowledge of community police officers. In the second workshop, emergency response officers participated as well. The insights from this workshop validated the choice of stimulating community policy officers and emergency police officers to share knowledge with each other: a suggestion to contact B is sent to A, while at the same time a notification is sent to B that A received this suggestion, after which they can begin communicating.

Fourthly, the refined service concept was developed into a prototype. The mock-up was then evaluated with police officers in a workshop, with the aim of discussing the functionality, followed by a small-scale test in which police officers used the mock-up on the street during their work.

It may be noted that various design choices were made during the four stages that led to a gradual shift in the objective and focus of the project. So far, the final step in the design of the service is the validation of the added value of the PolicePointer service. In a quantitative approach, an analysis was carried out of the amount of past incidents in which the service could have been useful. In a qualitative approach, workshops were conducted in which storylines involving the use of the service were rated by practitioners.

In terms of the dynamic STOF model that was presented in Chap. 2, the service developed from a vague concept to a service that might be implemented and rolled out. Up to that point, the design choices in the service component had been made in relative isolation from choices in the technology, organization and finance domains, i.e. there was no explicit balancing of choices in the four STOF domains. As a result, it was still uncertain whether the service concept could actually be implemented within the police organization.

The next step was, therefore, to design the business model. First of all, the organizational context of the police organization was analyzed extensively, by interviewing relevant stakeholders and studying existing research into service innovation at the police force. On the basis of this, the Critical Design Issues (CDIs) were addressed. Decisions that were already implicitly present in the service concept were made explicit. With regard to the remaining issues, specific options were recommended. Whenever there was uncertainty about what choice to make, the trade-off was explicated and further consideration by relevant stakeholders was recommended. In the remainder of the paper, we discuss these CDIs for the four domains.

9.2 Service Domain

The assumption underlying the PolicePointer service is that police officers often require specific information to do their job well. While a large part of this information is stored in information systems and while there are formalized procedures for sharing knowledge, officers possess a lot of additional, implicit knowledge. Unlike explicit information, which is stored in databases and can be retrieved through a mobile device, it is difficult to make implicit knowledge available and share it. How can a person know which colleague has what implicit knowledge? And which colleague needs that specific knowledge at a given moment? The PolicePointer is based on the idea that people can share implicit knowledge once they know who needs and who possesses certain knowledge.

An example of the use of the PolicePointer: Police officer Paul is being called for emergency assistance, as he has to attend a domestic violence incident. Community officer John has been at the same address last week for a similar incident. He made several observations, e.g. the next door neighbor can be helpful in solving the conflict. He also made specific arrangements with the residents. Some of these observations and arrangements have been stored in the information system, but some details have not been explicitly included. While he receives the emergency alert, the PolicePointer-enabled handheld device suggests Paul to call community officer John, including a short explanation as to why John may be relevant. In addition, John receives a notification that Paul may call him. Paul can then decide to call John for advice or John can call Paul to give him advice proactively.

The service uses an alert to query the information system for comparable incidents, e.g. incidents on the same address, involving the same person(s) or with similar key words. In this way the service can find relevant persons, i.e. police officers or social workers, who have been involved in earlier incidents and are likely to have additional information. On the PDA screen, a list of relevant persons is presented, including their whereabouts, a short note explaining why they could be relevant, and a phone number for direct contact. Figure 9.1 shows the design of the PolicePointer user interface.

Fig 9.1. Design of PolicePointer user interface

The we-centric element of the PolicePointer service concept raises several issues in the service domain with regard to *Creating Value Elements*. We-centric services are aimed at a group of users rather than individuals. The underlying assumption is that people are generally willing to help each other by sharing information. On some occasions one provides information to others, and on other occasions one receives information from a colleague. In the end, this benefits the group as a whole. In the case of the police force, this means that society is being served better as the quality of the work produced by the officers increases.

While the we-centric concept and the underlying notion of tit-for-tat may sound familiar to knowledge workers, it conflicts with the way of working within police organization. First of all, the division of labor within police departments is typically part of a hierarchical, top-down line of command. By contrast, the aim of the PolicePointer is to stimulate police officers on the street to take control into their hands and decide for themselves when and how to share information with others. As such, the service concept is at odds for the hierarchical steering mechanisms within police departments. This is an example of how user-centered design – or rather: participatory design (Schuler & Namioka, 1993) – begins at the work floor or on the frontline, and is designed to emancipate them, which may sometimes cause friction with existing (top-down) power structures.

Furthermore, emergency response officers and community officers typically work within different departments. In the Netherlands, police departments increasingly receive individual, performance-based funding, which is part of the new public management movement. However, as far as the PolicePointer is concerned, officers from one department are expected to help those from another department. To assure that the department managers will adopt the service, they should feel that the service does not jeopardize their performance objectives, i.e. there should be a balance between the time and effort invested in helping other departments and the benefits from being helped by those departments.

Directly related to this issue are the possible effects on performance in general. Adopting the PolicePointer is unlikely to lead directly to cost reductions, e.g. having fewer officers on the street. Because police forces have limited budgets, the PolicePointer should show significant, measurable quality improvements if the costs involved in adopting the service are to be justified.

Finally, there is also a risk of strategic behavior at the level of the individual user. While officers who provide assistance and information to others invest time and effort, those receiving help derive an immediate added value from the service. Obviously, officers should feel that these kinds of situations are more or less balanced, or they may refrain from helping other for strategic reasons.

Another issue in the service domain is *Targeting*. At some point in the project, the designers of the service considered supporting the collaboration between response officers, community officers and so-called network partners, e.g. firemen, school directors, concierges, et cetera. This could have increased the potential added value of the service, because there were likely to be more events in which the PolicePointer could have been used. However, it became clear that this would raise technology and legal issues, with information having to be shared across various information systems outside of the police domain. Also, the device should be accessible to all the network partners. Moreover, it would raise organizational issues, with a larger number of organizations becoming involved in the decision-making process. There was a clear trade-off between choosing a limited target group leading to reduced benefits from the service for end-users on the one hand, and choosing a broader target group leading to higher technological and organizational complexity.

Issues concerning *Branding*, *Trust* and *Customer Retention* are considered less prominent. Promotion and viral marketing are expected to be relevant in the adoption of the service by the officers. Especially when the service does not deliver the expected value, negative word of mouth may damage the image of the service, and damage user trust in the service.

9.3 Technology Domain

During the design of the service concept, few decisions regarding technology had been made. A highly important design issue within the police domain is *System Integration*. As we discussed in the introduction to this chapter, there is a wide variety of ICT systems being used within the Dutch police force. In fact, the core task of the central ICT organization is to standardize, unify and maintain the various applications. This is done using an ICT reference architecture that each new service should comply with before being implemented. As a result, it is important to integrate the service with existing information systems, and to assure a fit with the existing ICT architecture and standards.

System integration is also important from the end-user's point of view. Police officers have limited space on their belts, and they already carry around one or more mobile devices. This meant that the PolicePointer should run on existing mobile devices. In addition, end-users should not have to enter information into the PolicePointer system by hand, because that would require too much additional effort. Basically, the service should be technically integrated with existing mobile information services such as P-Info that provides secure mobile access to specific databases.

Security is a critical design issue for any service within the police domain. Information retrieved by police officers should not fall into the wrong hands for reasons of privacy, confidentiality of information and police officer safety. For example, a general rule is that information may not be stored on a handheld device, but may only be retrieved from an external server. In addition, there are specific rules regarding the process of criminal sanctioning that imply which type of police officer may and may not access and share information. The we-centric element adds a dimension of complexity to this last issue, because the service makes it possible to share information among various police departments. As a result, the risk of information being disclosed to an unauthorized police officer is much greater than it is for other mobile services. Consequently, when implementing the service, the rules for security levels relative to the role of the police officer should receive special attention.

9.4 Organization Domain

Service innovation within the Dutch police force is a challenge specifically when initiated by the central ICT organization. As we discussed in the previous section, the main task of the central ICT organization has to do with maintaining, standardizing and unifying existing applications. Some of the police regions we interviewed criticized this internal focus, claiming that it would constrain attention too much on maintaining existing systems rather than developing innovative solutions. Moreover, as mobile services are typically hard to maintain, police regions argued that this causes resistance on the part of the ICT organization with regard to developing such services. Another issue regarding service innovation is that application developers at the central ICT organization are not working at the same location as police officers, which complicates a user-centered design approach.

While service innovation may not be the core strength of the ICT organization, police regions have no choice but to involve them in the innovation process. Police regions wanting to develop new services are obliged to consult the central ICT organization first, because it is the preferred supplier. The ICT organization is in charge of transforming functional specifications into technical specifications, coordinating application development and integrating the application with existing systems. Moreover, all decisions regarding any new system have to be formally approved by the ICT organization. At the same time, the central ICT organization depends on the police regions, because the regions ultimately decide whether or not to adopt the service. This is crucially important because the central ICT organization becomes increasingly dependent upon funding based on actual adoption of services by police regions. In addition, testing new service concepts can only be done with officers at the police regions, which means that police regions and ICT organization have to work together in order to develop new services such as the PolicePointer.

To organize the collective activities of the central ICT organization and the police regions, *Network Governance* is a crucial issue. The ICT organization increasingly formalizes coordination during service development. All steps in the process are recorded in annual plans, requests for change and project plans. The police regions comment that this means that the central ICT organization focuses too much on procedures, instead of hands-on problem solving. In addition, it is simply too costly to invest time into these reports. It is also mentioned that, before developing an innovative service, it is very hard to put everything on paper due to uncertainty about what will happen during service development.

The formalized governance structure is enforced by a rigorous use of a reference architecture by the central ICT organization. New services should comply to this architecture before they can be developed. While the strict use of a reference architecture is understandable for services in a mature stage, it draws too much attention to technical details, rather than the intended user experience, when it comes to services that are still in a proof of concept phase and that are being developed and evaluated. It also limits the pragmatism required when developing an innovative concept whose end result is not yet clear. A stakeholder outside the ICT organization suggested that the reference architecture implies a major decision at the start of a new service development cycle. On the one hand when the central ICT organization is involved early on, this guarantees compliance with the reference architecture. However, it would also mean that the majority of the budget is being spent on complying with the centrally enforced architecture, although the service concept may not even be evaluated as worthwhile in the end. On the other hand, if the ICT organization only becomes involved when the service concept has been tested and prototypes and pilots have been developed, the costs may be even higher, because the concept and systems may have to be adapted to fit the architecture.

There is also the issue of trust, which is important in the service/concept development phase, because unforeseen events are likely to occur which cannot be accounted for in contracts. In addition, because changes in the original design of the service are likely to be made in this phase, strict contracts and project plans would have to be revised often. There is a general lack of trust between police regions and the central ICT organization, partly as a result of ongoing reorganizations that have led to uncertainty. While trust between the organizations is growing, some police regions comment that recent experiences in service innovation with the central ICT organization have damaged their existing relationships.

With regard to the PolicePointer case, a recommendation was made to complement contracts (i.e. service level agreements and project plans) with trust-based, close collaboration to benefit the innovation and implementation process. As we discussed above, most cooperation with the ICT organization is based on hierarchical structures and strict rules and contracts. Developing trust and an entrepreneurial approach based on partnership during the implementation and roll-out of the service was deemed more advisable.

Another important design issue is *Partner Selection*. As we discussed earlier, the central ICT organization cannot force the police regions to adopt a service. Therefore, involving the regions at an early stage is important to get feedback on the service concept and ensure commitment to the innovation process. So far, few police regions were directly involved in the development of the PolicePointer. Therefore, a crucial issue would be how to involve the police regions.

Also related to partner selection is the choice towards the technology providers. As the service concept and prototype were developed by the central ICT organization in collaboration with two technology providers, a choice had to be made whether to outsource the implementation of the system to these or other providers, or to insource it.

9.5 Finance Domain

The decisions in the finance domain depend on the trajectory chosen for the innovation process. In most cases a police region requests a new service from the central ICT organization. Once all regions agree on the need for the new service, the central ICT organization starts developing the service. After testing the service concept, a pilot is conducted within a specific region where the service actually runs on the infrastructure that will be used. After that, the service is implemented, integrated into existing systems, and rolled out on a nation-wide scale. An alternative, less common trajectory, is one where the central ICT organization auto-nomously decides to develop a new service, after developing a detailed business case. In this trajectory, the account managers of the ICT organization have to sell the service to the police regions after it has been developed. Investments in this case are made by the central ICT organization.

To implement the PolicePointer service concept in the police organization, a trajectory, in which police regions request for a new service, is not applicable, because the service concept has already been developed and validated. The other trajectory would be that a business case for the service is developed by the central ICT organization. A detailed design of the finance component of the business model would then be required to show the viability of the system in terms of costs and revenues. After implementing and rolling out the system, police regions could autonomously decide whether or not to adopt the service. This trajectory requires *Investments* to be made by the central ICT organization, and involves the *Risk* that police regions will not adopt the service. *Pricing* is therefore an important issue here, as the added value of the service would have to outweigh the costs involved in adopting the service by the police regions.

An alternative, more feasible business case trajectory would be to frame the service as part of a new release of existing mobile services such as P-Info, in which case, the PolicePointer would be bundled with these existing services. As payment would be included in the bundle tariff for the existing mobile services, *Pricing* can be determined by the central ICT organization. Although *Investments* are still made by the central ICT organization, there is less *Risk* because police regions would automatically adopt the service as part of the bundle. Choosing this trajectory would probably make the discussion on the business case for the service much simpler.

9.6 Conclusions

In retrospect, the PolicePointer project has been highly ambitious. It not only involves challenges relating to we-centric service elements, but also the challenge of service innovation within the highly unfavorable Dutch police organization. Overall, there are two design issues that appear to be the most critical in this case.

As a result of the we-centric elements of the service, the design issue *Creating Value Elements* is critical. The PolicePointer allows police officers to make their own decisions on how to share information, which is at odds with the hierarchical steering mechanisms within the police organization. In addition, there are risks of strategic behavior, as there should be a balance between the effort involved in helping others and the value received from being helped. Moreover, Dutch police departments are increasingly funded via individual, performance-based budgets, and the idea of helping other departments without guarantee of a direct gain for one's own department conflicts with this trend.

Within the police context, *Network Governance* is a critical design issue. The 26 autonomous police regions and the central ICT service organization are interdependent during the service innovation process. The typical governance mechanisms are highly formalized, as the central ICT organization focuses on procedures and complying with standards and reference architecture. Because there is little trust and close cooperation between the police regions and the ICT organization, service innovation is a lengthy and costly process. In case of the PolicePointer, a recommendation was made to govern the collective action of ICT organization and police regions based on a combination of contracts (i.e. service level agreements and project plans) and trust-based, close collaboration.

This chapter illustrates how the STOF model can help identify issues in the business model after a service concept has been developed and tested, and is ready to be implemented. We used the model as a 'checklist', i.e. to make implicit design choices explicit; to make recommendations on unresolved issues; and to identify which design choices still have to be resolved by relevant stakeholders.

Chapter 10 Balancing Customer and Network Value of Mobile Payment Services

E. Faber and H. Bouwman

The challenging aspect of business models is that they require managers to connect and balance various design choices and business model components. Mobile payment services are an interesting case because they have two types of customers, buyers – the people who pay – and merchants – the people who receive the payments. Mobile payment services require service providers to find a balance between the interests of buyers (or end-users), merchants, network providers and financial institutions In this chapter we illustrate how three different mobile payment service providers have tried to position themselves in an emerging mobile payment market. We focus on the connection between the design of the customer value and that of the value network.

10.1 Mobile Payment Services

Generally speaking, mobile devices are expected to play an important role in commercial transactions, such as payment services (see e.g. Krueger, 2001). Mobile payment services are payment services in which at least a part of the transaction (initiation, authorization execution, and/or confirmation) is carried out with the use of a mobile device, such as a PDA or mobile telephone, and of wireless or mobile network technologies like Bluetooth, 3G networks or Near Field Communication (NFC). In an international survey conducted among 5,600 mobile phone users, Kearny found that 46% of the respondents would use mobile payment services if they were to become widely available (Mobinet, 2002).

Given these high expectations, it is hardly surprising that over the past few years there have been many mobile payment initiatives. Most mobile payment services that were available in Europe in early 2002 have been discontinued, for instance Paybox and Simpay (Dahlberg, Mallat, Ondrus, & Zmijewska, 2007). However, the introduction of mobile payment services proceeds at a slow pace, not least due to the difficulty of developing feasible and viable business models. Any mobile payment service provider has to come up with a proposition that is interesting to both consumers and merchants. This dual focus, and the subsequent 'critical-mass problem' makes any market introduction more complicated (Markus, 1990; Oliver, Marwell, & Texeira, 1985). Moreover, businesses need to account for the business logic, not only of the financial services sector, but of the retail and mobile telecommunication sector as well. This creates a multi-party, cross-sector problem, the solution to which is fundamental to the development of mobile payment services.

Important questions in designing business models for mobile payment services are how to create sufficient customer value and who to involve in the value network. For instance, how indispensable are financial institutions or bank licenses issued by regulatory authorities in the banking industry when offering mobile payment services? Based on a literature survey (see Chap. 2), we focus on the connection between the customer value of (mobile payment) services and the value network by which they are offered. The customer value of service offerings and the value network needed to realize customer value are closely related, as can be seen in Fig. 10.1. An important research question is how the design of a value network influences the customer value of a service offering and vice versa.

We will explore this relationship by analyzing the business models of three mobile payment services.

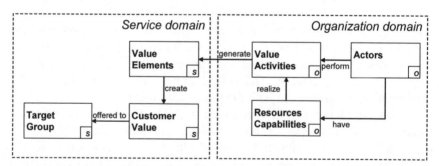

Fig. 10.1. Connection between customer value and value network

10.2 Research Method

The findings we discuss in this chapter are based on an exploration of three mobile payment services initiatives (Faber, Ballon, et al., 2003). Before investigating the actual cases, we consulted industry reports, academic literature and company web sites, which allowed us to arrive at an overview of mobile payment initiatives and to limit the scope of the case studies. The cases we selected are summarized in Table 10.1. They were selected because they represented different market introduction strategies. For every case, we consulted representatives from mobile payment providers, retailers and financial institutions. We conducted semi-structured interviews and talked informally with dozens of company representatives. Given the exploratory nature of the subject the interviews were semi-structured. The interviews were recorded and transcribed. The data from the interviews were supplemented with information from company websites, industry reports and academic literature. We made case descriptions which we used to conduct cross-case analyses. To ensure internal validity, all of the researchers involved used the same interview and case description templates. In addition, they conducted the interviews in various combinations and reviewed and discussed each others' case descriptions. The case descriptions were reviewed by the informants. The cases were then stored in a database to which all the researchers had access, for analytic purposes.

Table 10.1. Mobile payment cases

Case	Characterization
Mobipay	*Bank-oriented*: customer base of financial institutions is used to introduce mobile payment
Moxmo	*Independent*: customer base is built independently of financial institutions, existing payment infrastructure is bypassed
Mobile2Pay	*Independent*: customer base is build independently of financial institutions. Financial institution plays a role as trusted third party

Since both consumers and merchants play an important role in the distribution and acceptance of mobile payment products, the value proposition has been analyzed from the perspective of both these actors. Important value elements with regard to mobile payment are *Security,*

Trust, Ease of Use, Cost and reach, i.e. *Access* for consumers as well as merchants (Ondrus & Pigneur, 2006) (see Table 10.2). Ondrus and Pigneur also mention, for instance, flexibility – the degree to which a technology can be adapted to various types of payments – and maturity of technology.

Table 10.2. Mobile payment value elements

Value element	Definition
Security	*End-users and merchants*: degree to which the confidentiality and integrity of transaction information is protected
Ease of use	*End-users and merchants*: degree to which it is possible to operate a payment solution without having to overcome a steep learning curve
Cost	*End-users*: the costs (e.g. transaction, registration and communication costs) associated with using the mobile payment solution
	Merchants: the costs associated with supporting mobile payments (e.g. investments in transaction systems)
Reach	The number of points of sale at which customers can use the mobile payment solution
	The degree to which merchants can reach new customers

In the next sections, we describe the services, with a focus on the service and organization domains.

10.2.1 Mobipay

Mobipay started in December, 2000, initiated by the leading Spanish financial institutions Banco Bilbao Vizcaya Argentaria (BBVA) and Santander Central Hispano (SCH), as well as all the Spanish mobile telephone operators (Telefónica Móviles, Vodafone and Amena). The objective was to develop and promote an international mobile payment solution based on a co-operative model involving mobile operators and financial institutions. Mobipay International is the holding company that owns the local initiatives in various countries. Mobipay was terminated in 2004.

Mobipay's value proposition was to provide mobile and single access to the existing payment infrastructure. It provided a transaction platform that was capable of supporting all kinds of payment methods. It allowed consumers to pay through existing and trusted electronic payment methods, using their private mobile phone as an authentication terminal. Mobipay focused primarily on the financial institutions and telecom operators. Although it provided banks and telecom operators with reasons why mobile payments add value to end-users, it failed to offer a value proposition directly to these end-users. In Spain, the main selling point with regard to financial institutions was that Mobipay was expected to increase intermediated payments.

Mobipay provided a mobile transaction platform to financial institutions that could be used with existing payment methods. The mobile operators were more or less part of the distributed platform through the integration of their mobile access services into the platform. The mobile network operator provided mobile communication services to Mobipay, which was used to access the payers' verification terminal. Mobipay offered financial institutions single and mobile access to the existing payment infrastructure. As far as Mobipay was concerned, ensuring the support of financial institutions was important because of their relationship with customers and retailers, who were the users and acceptors of mobile payment solutions. Mobipay received a fee per transaction for its services. The financial institutions handled the payment of the transactions, while Mobipay was responsible for the authorization, authentication, electronic assembly and delivery of payment transactions. Retailers were responsible for providing payment methods to consumers, and in return they would be adequately rewarded for the products and services. The payment services were provided by the financial institutions (or third party payment providers) and were paid for by the retailers. Table 10.3 summarizes the design choices for the critical design issues.

The Mobipay project was also promoted internationally. Although the website still exists, there have been no activities since 2004. Vodafone, a company that was involved in MobiPay, was also involved in Simpay, a competing mobile payment service that was also terminated in 2005, due to the fact that one of the mobile telecom providers decided to launch its own mobile payment system.

Table 10.3. Design choices for Mobipay's mobile payment services

Target group	Value elements	Offered customer value	Resources and capabilities	Actors
Consumers	Security	Secure payment	Secure authentication and authorization	Mobipay
		Control over transactions	Transaction management	Financial institutions
		Trusted Third Party	Institutional rules of conduct	
	Ease of use	Mobile phone as access device to existing payment infrastructure	System integration	Mobipay
	Cost	Not different from normal transactions	Cost efficient payment infrastructure	Financial institutions
	Reach	Covers a wide variety of transaction situations and is supported widely in Spain	Customer base	Financial institutions and mobile operators
Merchants	Security	Secure payment	Secure authentication and authorization	Mobipay
		Trusted Third Party	Institutional rules of conduct	Financial institutions
	Ease of use	No changes needed in existing points of sale	System integration	Mobipay
	Cost	Cheap electronic point of sale (Spain)	Cost-efficient payment infrastructure	Financial institutions
	Reach	Increased customer reach	Access to customer base	Financial institutions and mobile operators

10.2.2 Moxmo

Moxmo was a Dutch payment initiative providing mobile payment solutions to merchants, primarily on the Internet and in non-POS situations. The company was founded in 2001 and, together with Global

Payways, went bankrupt in 2004. Moxmo was an interesting mobile payment initiative, being one of the few solutions that was set-up independently of banks and (mobile) telecom operators. Table 10.4 summarizes the design choices for the critical design issues.

Moxmo was a mobile wallet service that allowed consumers to make secure and direct payments to anyone who owned a mobile phone or who accepted Moxmo as a payment method. Since Moxmo did not regard mobile payments as product in itself, it worked closely together with service providers in the development of new innovative service concepts that could incorporate mobile payment. This was seen as an important prerequisite for the further growth of mobile payments, and as such of Moxmo in particular. Moxmo's value proposition with regard to consumers was that it offered convenient (any time any place) and secure electronic payments. Moxmo focused on the micro-payment market segment, in particular person-to-person payments, Internet payments, topping-up of prepaid accounts and ticketing. At a later stage Moxmo intended to extend these services to include parking, international transfers, debit card payments, customer cards and, ultimately, payments in stores.

Moxmo was a start-up company owned by Global Payways. Global Payways played three roles in Moxmo, divided into distinct business units. First of all, it was involved in developing service propositions in co-operation with third parties. These propositions should have resulted in high-end value services and products involving the services of Moxmo. Secondly, Global Payways processed the transactions generated by Moxmo. Thirdly, it wanted to control the deposits stored on the electronic wallets of its customers. To do this, Moxmo needed to acquire a license as an Electronic Money Institution (EMI). As far as we know, between 2001 and 2004, Moxmo was continually in the process of obtaining such a license. Meanwhile, customer deposits were controlled by the Dutch bank ABN-AMRO. Retailers were supposed to play a role as acceptors and distributors of Moxmo. In their capacity as distributors, retailers were expected to actively promote the Moxmo payment method to their consumers. Moxmo offered revenue sharing for each payment, hoping that that would win over consumers and merchants, which in turn would help the company build its brand. The operational management of the transaction platform and mobile wallet administrator was outsourced to an Application Service Provider. Finally, mobile operators facilitated the mobile access between the users' mobile phone and the Moxmo transaction platform.

Table 10.4. Design choices for Moxmo's mobile payment services

Target group	Value elements	Offered customer value	Resources and capabilities	Actors
Consumers	Security	Secure payment	Secure authentication and authorization	Moxmo
		Control over transactions	Transaction management	ABN-AMRO
	Ease of use	Mobile phone as prepaid wallet	License for EMI	Moxmo
	Cost	Low cost through bypassing of existing payment infrastructure	Cost-efficient independent payment infrastructure	Moxmo
	Reach	Increasing number of retailers	Access to customer base	Moxmo and retailers
Merchants	Security	Secure payment	Secure authentication and authorization	Moxmo
		Guaranteed payment	Risk management?	Moxmo
	Ease of use	–	–	–
	Cost	Low cost through bypassing of existing payment infrastructure Revenue sharing	Cost-efficient independent payment infrastructure	Moxmo
	Reach	Increased customer reach	Access to customer base	Moxmo and Telecom operators

10.2.3 Mobile2pay

Mobile2Pay is a mobile payment initiative that was launched by Smart Concepts in October 2002. Its main objective was to create an interactive mobile sales channel that would enable the payment and delivery of goods, in particular impulse purchases. It would appear that the service ceased all operations since Mid-2007.

Mobile2pay formulated its value proposition as 'seeing is having'. Consumers could respond directly to advertisements published in magazines or broadcast on the radio, using their mobile phone as a transaction device. The main advantages of Mobile2pay as far as consumers were concerned were assumed to be speed of use (impulse buying and fast processing), ease of use (small smart device) and benefits (discounts). Mobile2pay focused on retailers, and its main objective was to set up a mobile sales channel, rather than to provide a mobile payment system. Based on this idea, Mobile2Pay defined a strategy for mobile commerce in general, and used mobile payments as an enabling functionality. Mobile2pay focused on medium-sized and macro-payments. Table 10.5 summarizes the design choices with regard to the critical design issues.

Table 10.5. Design choices for Mobile2pay's mobile payment services

Target group	Value elements	Offered customer value	Resources and capabilities	Actors
Consumers	Security	Secure payment	Secure authentication and authorization	Mobile2pay
		Control over transactions	Transaction management	Mobile2pay
		Guaranteed delivery	Escrow service	Fortis bank
	Ease of use	Mobile device as debit card	Automatic collection	
	Cost	Price reductions when ordered with mobile phone	Access to customer base	Retailers
	Reach	Increasing number of retailers	Access to customer base	Retailers
Merchants	Security	Secure payment	Secure authentication and	Mobile2pay
		Guaranteed payment	authorization	Fortis
			Escrow service Dynamic spending limit for users	Mobile2pay
	Ease of use	–	–	–
	Cost	Increased sales through impulse buying	Anytime and anywhere payment	Mobile2pay
	Reach	Increased customer reach	Access to customer base	Mobile2pay and mobile operators

Retailers were expected to provide payment methods to their customers as an added service. Fortis bank acted as trusted third party with regard to the payment transactions between consumers and retailers, receiving an authorized and complete payment transaction from Mobile2Pay, and returning a bank guarantee based on the consumers' creditworthiness. After receiving the guarantee the retailer could ship the goods. After a retailer was informed that the goods had been received by the customer in question, the bank transferred the money to the retailer's account. By using a dynamic spending limit, Mobile2pay filtered out defaulters. Customers were rewarded for keeping their promises (faithful payment) by an increased spending limit.

Mobile2Pay handled the consumer authentication and the authorization of the payment process. Finally, mobile operators facilitated the mobile access between the users' mobile phone and Mobile2Pay transaction platform.

10.2.4 Cross Case

The mobile payment providers we examined in this study used different strategies to obtain critical mass, but apparently none of them succeeded. Whereas Mobipay focused on bringing together financial institutions and telecom operators, Moxmo and Mobile2pay directly tried to convince merchants and customers of the added value of mobile payments. However, customer groups had different needs and wishes.

Security. All the mobile payment providers discussed in this chapter were aware that security and trust were of the utmost importance. However, they used different mechanisms to generate security and trust. In the case of Mobipay, consumers dealt with their trusted home banks. Mobipay expected to benefit from the existing relationship between the banks and their customers. Moxmo, on the other hand, had to prove its trustworthiness to consumers by recurrent positive experiences. The company promised guaranteed payment to merchants. Mobile2Pay, finally, used an escrow service to eliminate the risks facing consumers and merchants, and was thus able to guarantee product delivery to consumers and payment to merchants. Moreover, by deploying a dynamic spending limit, Mobile2pay filtered out defaulters.

Ease of Use. The payment initiatives discussed in this chapter varied with regard to the entry barriers for consumers. In the case of Mobipay, consumers did not need to register at all, provided that they had access to a bank account, while in the case of Moxmo, users needed to register and open a new bank account, and with Mobile2Pay, they needed to register and authorize an automatic direct debit. As far as merchants were concerned, integration with existing payment products was an important issue. They were not all that eager to implement a new payment product in addition to existing payment products like debit and credit card payments. Only Mobipay presented a convincing case with respect to this value element.

Cost. Mobipay focused primarily on financial institutions (banks and payment brands), and to a lesser extent on telecom operators. It left the promotion of mobile payment to the financial institutions and telecom operators. Moxmo saw mobile payments not as a product in themselves, and worked closely together with service providers to develop new innovative service concepts in which mobile payments could play a role. Moxmo offered revenue sharing and new service concepts to merchants. Consumers were offered a convenient and secure payment service. These value proposition elements could hardly be regarded as 'deal clinchers', but rather as 'dissatisfiers' (their absence provides a negative experience). Mobile2pay tried to persuade consumers to use its services by offering price reductions on products that were paid for via Mobile2Pay. The impulsive nature of purchases was emphasized (seeing is having) and promoted as a valuable experience to consumers. Mobile2pay offered merchants an interactive transaction channel, in addition to the Internet and physical points of sale.

Reach. Mobile payment providers used different strategies to get merchants and consumers on board. Mobipay relied on financial institutions to convince merchants and consumers of the value of mobile payments. Moxmo and Mobile2pay could not rely on the customer base of financial institutions, and had to persuade customers themselves.

Merchants seemed to be willing to adopt mobile payments if they resolved some of the problems associated with existing payment products. Guaranteed payment in particular was valued highly. This could be realized in different ways, as illustrated by the Moxmo and Mobile2Pay cases.

Consumers did not seem to view mobile payments a service, but rather as 'a necessary evil'. Consequently, low transaction fees, ease of use and guaranteed delivery were 'dissatisfiers' rather than 'deal clinchers'. For mobile payment providers this meant that it was important not to promote mobile payments as a product in themselves, but as an enabler of new value adding services. Although the two independent initiatives provided interesting examples of this concept, ultimately they proved unsuccessful.

Although both Moxmo and Mobile2pay were marketed as independent initiatives, our findings indicate that financial institutions did play an important role in both initiatives. Because Moxmo did not have a banking license, it had to rely on ABN-AMRO for the management of wallet deposits. By using escrow services, Mobile2Pay was able to guarantee payments to merchants as well as delivery to consumers. By including financial institutions in the value network, the mobile payment providers were able to reduce the risks facing merchants (guaranteed payment) as well as consumers (guaranteed delivery). However, this inevitably led to higher transaction costs. Moreover, with hindsight we suspect that the reasons the banks participated were primarily strategic in nature.

An important decision involved the question whether or not to include financial institutions in the value network. Required guarantees and transaction costs are important influence factors. For micro payments, which require fewer guarantees and low transaction costs, it would seem to make little sense to include financial institutions. It makes more sense to do so in the case of medium-sized and macro-payments. However, the question whether mobile payments that are limited to micro-payments are attractive enough for consumers is open to debate. The cases discussed in this chapter reveal that certain value elements can be realized in different ways and that, depending on the target group, it is possible to bypass dominant actors, for instance financial institutions, in the value network. Nevertheless, financial institutions have a strong position and by-passing them is problematic, as is indicated by the bankruptcy of Moxmo, and by the fact that the other two services did not manage to survive either.

10.3 Conclusion

Based on the three cases discussed in this chapter we can draw initial conclusions with regard to customer value. First of all, it is important to realize that services that are innovative in technological terms are not necessarily perceived as very innovative by their intended customers. Mobile payment services in themselves offer yet another way, in addition to cash money, credit and debit cards, of taking care of payments. This implies that positioning a mobile payment channel next to already existing payments channels is problematic. Although such a service may be new to the world from a technological point of new, that does not mean that customers are guaranteed to see the added value for them. Furthermore, merchants, who after all control the relationship between the mobile payment services and the end user (the customer), are unlikely to adopt the mobile payment channel if it does not solve some of their problems. This means that, in the service offering, the issue of value has to be addressed very carefully for both customers and merchants. A possible alternative is to make mobile payments an integral part of other innovative mobile services. The mobile payment service can be presented as an additional feature of such a new service.

With regard to the value network, we see that by-passing financial institutions is rather problematic. In light of their powerful position within the value network, it is almost impossible to by-pass banks, especially in the case of service that will not be limited to micro-payments.

As far as our conceptualization of the design of businesses models is concerned, it is clear that looking at the value of a service from a customer perspective is not enough, but that we also need to take the perspective of the service provider into account. The cases discussed in this chapter show that mobile payment providers had to position their services in a relationship between financial institutions and their customers, were there are many alternatives and where the dominant role of players in the existing value networks leaves ample space for new entrants (see also Zmijewska, 2007). In Chap. 13, we will see that in developing countries, were banks have a less prominent position, the development of mobile banking services is more feasible.

Chapter 11 Robustness of IPTV Business Models

H. Bouwman, M. Zhengjia, P. Van Der Duin, and S. Limonard

The final stage in the STOF method is an evaluation of the robustness of the design, for which the method provides some guidelines. For many innovative services, the future holds numerous uncertainties, which makes evaluating the robustness of a business model a difficult task. In this chapter, we apply scenario analysis to assess business model robustness, using digital television services over IP (IPTV) as an example. Before discussing the scenario analysis, we take a look at the discussion surrounding the IP television.

11.1 IPTV: Multicast Digital Television Over IP

IPTV is a broadcast or on-demand video service that uses the Internet Protocol (IP), which is streamed to a set-top box that can be connected to a PC or a TV-set. Today, IPTV is technically and commercially applicable. IPTV services are typically considered by telecom companies facing fierce competition from cable companies that offer triple play bundling packages (*Alcatel*, 2007) However, it is not yet clear whether or not new IPTV business models will be viable. Although there are many analysts who believe that the huge potential of the IPTV market will offer telecom operators opportunities to develop this new market, there are other experts who doubt the telecom operators' ability to compete successfully with cable companies, which in many countries possess significant market power in the TV market. In the current context, in which cable companies move toward data services and voice communication, a clear and comprehensive business model for IPTV is critically important to telecom operators. However, no template for a successful IPTV business model is available. Different players in the market emphasize different elements. Doherty et al. (2004), for example, focus on the right IPTV architecture and quality of services, while Liu (2006) is more interested in the financial prospects of the new IPTV model. As yet, there is no common and shared conception of business models.

In this chapter, we used the STOF method to develop business models for IPTV. We draw a distinction between the exploration phase and the exploitation phase. With regard to the exploration phase, we discuss design issues based on the results of experiments and market explorations by IPTV developers that are currently taking place. After discussing the trade-offs between the various design choices in the IPTV domain, we use scenario analysis to discuss issues that may be relevant in future exploration phases: IPTV developers will focus on issues like service (bundles) and more efficient marketing strategies. We use the scenario analysis to assess the robustness of the business models, like those of IPTV, that are characterized by high levels of uncertainty when they enter the exploitation phase.

11.2 The IPTV Business Model

In this section, we discuss some of the business model design issues regarding IPTV in the exploration phase, and the tradeoffs and relationships between relevant design options. The design space is defined by technological and market-related drivers.

Technological drivers for IPTV are (1) an increase in effective distribution capacity, aimed specifically at 'the last mile'; (2) an increased ability to process user feedback via recently developed, innovative technologies that increase the feasibility of interaction from the home environment; (3) an increase in storage and processing capacity controlled by viewers; and (4) a separation between applications from transport, to ensure quicker innovations in the application layer, regardless of transmission bottlenecks (Katz, 2002).

Market-related drivers and conditions include (1) market demand, e.g. thus far, Western Europe has proven to be one of the world's most vibrant markets with regard to IPTV, for instance MalIgne TV in France, FASTWEB in Italy and Telefonica's Imagenio in Spain; and (2) a convergence of information, telecommunication and TV industry that leads to fiercer competition. Technological and market-related developments set the conditions under which business models for IPTV have to be developed. We will address the various design issues in greater detail in our discussion of the four domains of our model.

11.2.1 Service Domain

Many telecom operators position IPTV as a service that directly competes on the mass-market with cable and satellite companies. Telecom operators expect they will be able to take advantage of their expertise and of consumer loyalty in the telephone and broadband (DSL) market. They have to leverage their resources (deep pockets) to increase the competitiveness of IPTV. There are three ways in which they can do so: (1) bundling video service with broadband and telephone service, (2) offering more value-added services (e.g. interactive services, Video on Demand (VoD), and Personal Video Recorder (PVR)), and (3) service portfolio focusing on exclusivity or on a wide range of 'long tail' niche channels.

11.2.2 Technology Domain

Within the technology domain, there are three layers of architecture when it comes to IPTV. The transport layer design has to ensure that the infrastructures meet transmission requirements. The design of the middleware and content layers has to be given careful attention, because the design requirements are vital to specific types of services (e.g. VoD, PVR, or focused service portfolio's) and to the interactivity of the services. The conditional access components and middleware control the access authorization, manage the Customer Relation Management (CRM) and take care of the encryption of the video program. The set-top box is used to decode programs on the end-user's side. Different IPTV providers opt in favor of different video application components, based on service design demands and financial considerations.

In the technology domain, bandwidth limitations, quality of service and level of middleware capability are the main issues. The telecom operators can choose different design approaches, either by adding large capacity immediately before launching IPTV or by doing so gradually. The former approach usually leads to upgrades of large portions of the infrastructure towards fiber optic cables, while the latter approach involves the plan to begin by updating most of the distribution network towards ADSL2+ standard. The choice with which telecom operators are faced is between the more advanced technology platforms and larger investment on the one hand, and a more evolutionary approach on the other (Fijnvandraat & Bouwman, 2006). Because video content is sensitive to signal quality, service quality is the second issue at stake. Service quality can be improved by, among other things, better transmission infrastructure and encoding/decoding equipment, all of which requires higher investment. An alternative design choice is to offer different service levels to different customers. Middleware capabilities, particularly for telecom operators providing broadcasting, VoD and value-added services, are crucially important with regard to integrating platforms that enable interactivity, conditional access and customer relationship management.

11.2.3 Organization Domain

Actors in the value network of IPTV include telecom operators, content, telecom equipment, and middleware providers, advertisers and consumers. Telecom operators participate in and organize the entire value delivery process. They produce or purchase the video content, and maintain the physical network and the hardware and software components of video application supported by equipment manufacturers. A content aggregator intermediates between content providers and the telecom provider. Advertisers can also be involved. Figure 11.1 presents an overview of the actors, functions and services involved in IPTV.

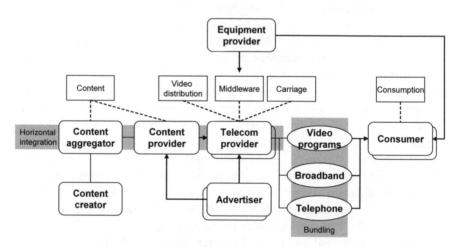

Fig. 11.1. Value network of IPTV

Ownership of and control over successive stages of the value 'chain' plays a decisive role in the way the IPTV services are marketed, i.e. bundled with other services (e.g. broadband and telephone) or separately. Another relevant issue has to do with the extent to which the IPTV provider (i.e. telecom operator) wants to be involved in the production of content.

11.2.4 Finance Domain

Technical and service design choices determine the costs of IPTV projects. For example, a high level of availability of services requires a large investment in setting up redundant servers to make sure there are sufficient video sources at peak times. Interactive services require financial investments to upgrade network equipment and to solve technical problems. The revenue models depend on revenue sources, for instance advertisement or flat rate models, and the accompanying pricing models. Revenue can then be calculated by the size of market and the expected market share.

In the current IPTV market, most telecom operators use a flat rate subscription model, and they are slowly adopting pay-per-view models as well as advertising in the case of VoD services. However, if more customized and personal services were to come available, a flexible pricing model would be more desirable and feasible in the future. To assess the robustness of the various designs we will use scenario analysis.

11.3 Scenario Analysis

Scenario analysis is an effective method of reducing uncertainties that helps firms formulate flexible and sustainable strategies and policies. As a tool for hedging future risk, scenario analysis is different from the more quantitative prediction methods. Instead of offering predictions, scenario analysis explores the future direction in a diverging perspective. Scenarios produce various possible developments and display a keener awareness of the uncertainty of trends (Bouwman & Van Der Duin, 2003). Scenario analysis focuses on understanding, capturing and describing possible future development rather than predicting the future based on a few selected variables. In the case of IPTV business models, scenario analysis will be used to identify the uncertainties that have a potentially major impact on the viability and feasibility of the business models during the exploitation phase (Limonard, 2006). Uncertainties often include elements that are difficult to quantify and model (Fijnvandraat & Bouwman, 2006). The scenario analysis helps IPTV operators develop a clearer idea with regard to the sustainability and feasibility of their business model.

The scenario analysis uses four reference scenarios that focus on the telecommunication industry and have a similar time horizon (2010), i.e. Drop, Dijkhuis, Van Der Duin, and Stavleu (2000), *European Commission* (2005), Foster Daymon and Tewungwa (2002), and Van Der Duin (2000). Based on these scenarios, we identified consumer attitudes and regulation related to industry structure as the most important uncertainties, and we will use them in the scenario analysis. We draw a distinction between sophisticated and regular consumers. Sophisticated consumers are fully aware of the advantage of new media products and services. They want to pay a reasonable price for customized and personalized services. Regular consumers are more conservative when it comes to new products. They are satisfied with conventional TV, prefer easy configuration of services and are price-sensitive. Bundling services at a discount is very popular among this group of customers. The industry structure dimension, as regulated by National Regulatory Authorities, on the one hand promotes a hands-off approach on the part of the regulators and domination of the market by telecom operators in the voice and broadband market. On the other hand, regulators play the role of interventionist and curtail the actions of the dominant market players. As a result, there are many players in the voice, broadband and television market. These two dimensions result in four scenarios, which are summarized in Fig. 11.2.

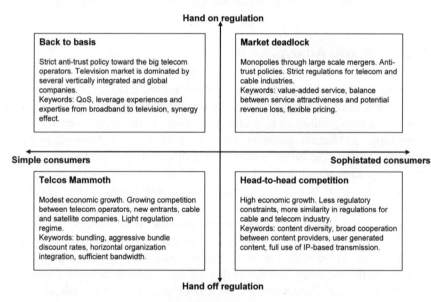

Fig. 11.2. Scenario dimensions, scenarios and critical issues

We will discuss the four scenarios in greater detail and illustrate the kind of design choices that become relevant in the exploitation phase of IPTV business models.

11.3.1 Telcos Mammoth

Telecom operators are not restricted by tariff regulation and have the freedom to form alliances with content providers. The younger generations of consumers are quick to adopt the new services that are offered at an affordable price. Most of them like easily configurable digital TV, telephone and broadband services. Consumers show an interest in bundling. The most popular packages are the discount offers from the multi-services providers. Bundling complementary products is an attractive strategy, given the fact that consumers will value bundled services more than the total added value of the separate elements. Telecom operators are the only players capable of providing quad-play. Although cable companies are triple-play veterans, they have a hard time catching up with telecom operators. However, market success is not simple in a market where several competitors (cable) are capable of offering similar service bundles. Telecom operators should focus on the preference 'simple' users have for easy and quick configuration and entice more consumers by introducing bundled products. Telecom operators are in a position where they can bundle the IPTV and broadband services. In most cases, IPTV and broadband services are independent products and the costs involved in providing them can only be reduced through bundling. However, in some cases they become complementary, which means that they add value for end-users.

Bundling has an effect on several business model aspects. In the technical domain, bandwidth is closely related to service bundling. Bundling broadband and video service requires high bandwidth capacity (e.g. video service requires at least 2–3 Mbps and broadband requires around 1 Mbps; the total is about 3–4 Mbps). Bandwidth will be the key enabler of bundling services. This means that a network upgrade to fiber or hybrid type is needed. In the organizational domain, horizontal integration is necessary in order to offer service bundles. It is a challenge for telecom operators to coordinate the activities of the newly established video business unit and the traditional broadband and voice business units, to provide seamless bundles and efficient billing systems. In the financial domain, it is very important to choose the right pricing and discount rate. Telecom operators have to balance the technical and financial design to control costs and to be able to offer a modest monthly rate to most consumers. This pricing strategy can be modified to predatory price levels that are enabled by economies of scale and scope. This can create competitive advantage and entice more consumers to the convenient triple play configuration.

11.3.2 Head-to-Head Competition

The regulatory framework leads to fair competition among equally strong players creating an expanding digital TV market. Different players compete at a national level and gain fair access to the content market. Competition facilitates innovation and strengthens the balance sheet of the digital TV industry as a whole. The market value of digital TV companies keeps increasing. Consumers are aware of the benefits of digitalization and experienced in selecting digital TV offers and programs. They prefer new services at an affordable price and are less interested in selecting services from a whole range of options. Consumers like to create and share content. The best-performing digital TV provider is the one that caters to the personal tastes of consumers and enables them to create, publish and share personal stories.

Sophisticated consumers search for the programs they prefer. Because of the personal preference among consumers, the content market becomes fragmented. Even the most popular programs do not attract much attention, so the result is an increase in narrowcasting: a large number of programs with small audiences, emulating a 'long tail' approach for television (Limonard & Tee, 2007).

Content diversity is of key importance. The success of a business model depends on the question whether or not operators can cater to individual preferences and capture the value located in the "long tail". In order to do so, operators have to look for extensive cooperation with various content producers, including content generated by viewers, and aggregators, to obtain as many and as diverse programs as possible, as well as to enhance their own capabilities to produce exclusive content (organizational domain). In the technical domain, IP-based distribution technology provides a competitive advantage, because it meets the requirements inherent in the 'long tail' to become economically interesting (Brynjolffson, Yu, & Smith, 2006). Because the IP infrastructure enables interaction and makes it possible to capture detailed information regarding user behavior, a powerful asset is created, giving telecom operators the opportunity of becoming an indispensable interface between customers, content producers and advertisers. Therefore, telecom operators should make full use of the advantages offered by IP-based transmission. Delivering customized service requires additional investments in content production and distribution. Although, according to Noam (2002), cable and Internet TV operators face similar costs in content production, the distribution costs involved in operating an IP platform are considerably higher. Individualization requires significantly larger transmission resources, which implies higher levels of investment.

11.3.3 Market Deadlock

Telecom operators have to go through numerous procedures to receive TV franchise permits and are actually blocked from engaging in content integration. Cable and satellite companies take advantage of this regulatory environment. They consolidate their business, and become dominant and vertically integrated national players. In this scenario, consumers already have quite a lot of experience with digital TV. They are interested in new services.

It is hard to generate profit for telecom operators due to a stagnating market. The most important CDI involves providing new value-added services. To achieve satisfactory penetration rates for IPTV service, telecom operators need to create a competitive edge by differentiating their services. These services are not exploiting the long tail for content, but focus on value-added services, for instance by providing PVR (Personal Video Recorder) and interactive services (TV email, TV internet and games). Choosing this strategy inevitably leads to conflicts with content providers. Content providers may decide to implement some form of IPR-protection, to prevent use of PVR. However, according to Katz (2002) history suggests that these measures will be overcome and consumers will have the capabilities to copy content and edit programs. This threatens the ability of content packagers to rely on the sale of advertising (Loebbecke & Radtke, 2005).

Subscription fees and advertisements provide the key revenues for television distributors. Value-added services, like PVR offered by telecom operators lead to revenue loss for television distributors, which means they are forced to take counter-measures. One option is to give extra rewards to consumers who are willing to watch advertisements.

In an IP-based network, it is possible to monitor consumer behavior, and telecom operators can use these data to grant monitory compensation or conditional access to premier contents (Katz, 2002). Telecom operators can also tolerate consumers copying contents and skipping advertisement if the value-added services can attract enough new subscribers. Actually, according to Mogg (2004) and Myers (2002), these two features are the biggest selling points as far as PVR is concerned. Telecom operators should make the decision based on the balance between customer value and economic benefits.

11.3.4 Back to Basis

Cable and satellite companies take advantage of the regulatory environment and consolidate their business. Most consumers are satisfied with existing television services provided by the cable and satellite companies. They are extremely hesitant to adopt new services and there is little demand for interactive services and Internet TV. Consumers decided to use bundling services because they are more convenient and easier to configure.

In this scenario, regulatory barriers and consumer indifference make it hard for telecom operators to offer IPTV service. There are few new opportunities for telecom operators to develop the kind of IPTV-services defined in the exploration phase of a successful service. Although bundling may still be an effective strategy, it will be difficult to convince consumers to switch from cable companies towards telecom operators. Market potential is limited. The content supply side is dominated by several vertically integrated global companies. It is difficult for telecom operators to enter the content market and compete with content providers.

The more realistic option available to telecom operators is to make better use of the existing technologies, services and resources, to support IPTV development. First of all, telecom operators can create a fresh image as new entrants, providing a high level of Quality of Service and service availability and excellent customer service. Better program quality and customer service are compelling reasons for consumers to move away from cable and satellite providers. Secondly, telecom operators should leverage their customer base in the broadband and telephony market. Thirdly, they can create complementary programs and content on the television and broadband platforms. However, the back to basis scenario does not include a dominant strategy for telecom operators.

11.4 Conclusion

Since the future of IPTV is uncertain, business models should be easily adaptable to changing external factors and uncertainties. In this chapter, we have used scenarios to assess the robustness of IPTV business model design choices. The scenarios represent various possible futures, in terms of the regulatory environment, industry structure and consumers' attitude towards IPTV service. By choosing the appropriate business model, telecom operators can sustain market competition and deliver customer value and economic benefits. Because the resources are limited, telecom operators have to focus on the major design choices in each of the scenarios when balancing the requirements of IPTV business model design. Table 11.1 summarizes the design choices for the four scenarios.

Table 11.1. Summary of major design choices for each scenario

	Scenarios			
Domain	Telcos Mammoth	Head-to-head competition	Market deadlock	Back to basis
Service	Bundling IPTV with other services	Content diversity	Value-added service	Bundling
Technology	Sufficient bandwidth	Middleware capability	Middleware capability	QoS
Organization	Horizontal integration	Partial vertical integration to enable cooperation with content producers (professionals and User Generated Content)	Partnering/ partial integration with service providers	Consolidate
Finance	Aggressive discount rate	Balance increased costs with increased revenues	Balance content related loss in revenues and extra revenue from advertisers and price differentiation	Consolidate, focus on internal synergy effect in network

The design issues we have identified as being of critical importance in each scenario do not necessarily undermine the importance of other design issues. In fact, telecom operators always have to balance design choices to deliver both customer value and economic benefits. For example, the content diversity design is closely related to the financial domain, because the costs involved in producing tailored personal services are far from trivial. It is important to find ways to control costs (e.g. broadcasting user generated content is one way to reduce cost). Providing value-added services means running the risk of losing subscribers and advertisement revenues, which makes the business model vulnerable if telecom operators fail to attract enough new customers. Telecom operators have to decide the extent to which they want to offer the new services based on market demand and costs.

In this chapter, we have illustrated how business model analysis based on the STOF model can be complemented with scenario analysis. Scenario analysis helps identify various future possibilities and provides a platform for analyzing decisions with regard to the design choices that are likely to play a significant role in an uncertain future environment. The competing views regarding future developments are helpful in that they make it possible to reduce the future uncertainties with regard to the viability and feasibility of IPTV business models.

Chapter 12 Mobile Service Bundles

T. Haaker

In this chapter we consider business models for mobile service *bundles*. The focus of the STOF model and method is on mobile services. Service bundling will be an important driver for the use of 3G+ services. Bundles of services may provide a 'bundle of benefits' to users, potentially increasing customer value. Bundling may also lead to cost efficiencies for providers coming from economies of scale or scope. Any decision regarding a bundled offering of services requires careful balancing of customer and network value.

Notwithstanding the wide application of bundling strategies still little is known about what constitutes a viable bundle, i.e. a bundle that creates added value for users and/or providers. Clear guidelines neither for the design of service bundles, nor for the underlying business models are available. In this chapter we use the STOF model to discuss Critical Design Issues (CDIs) for service bundling that influence customer value. The study is based on a literature review and a number of case studies of existing mobile service bundles. We use the results to extend the causal model for creating customer value with additional CDIs related to service bundling.

12.1 Service Bundling

Bundling of products and services is a well-known practice in marketing (Kottler, 1999). Bundling can be defined as 'selling two or more products in a package for a special price'. While the concept of bundling is in itself straightforward, any question regarding what products or services should be bundled, and under what circumstances, is generally hard to answer. Companies use bundling to create price discrimination (Adams & Yellen, 1976), increase sales (Venkatesh & Kamakura, 2003), reduce costs (Shapiro & Varian, 1999), or create entry barriers (Nalebuff, 2004).

In the telecommunication industry the bundling of services with handheld devices is common practice. The aim of the mobile operators in providing an almost-for-free device is to attract customers. The actual revenues should come from complementary service subscriptions and usage fees. Bundling of telecommunication services is generally limited to packages of communication services, e.g., combing voice and SMS, and triple or quadruple play offerings.

Bundling of mobile services is, however, not very common. With 'mobile services' we refer to all kinds of services that combine technologies and concepts from the domains of telecommunication, information technology, and consumer electronics. Service bundles are expected to play a central role in the success of next generation mobile services. Given the lack of a single killer application for mobile services, and that the mobile device and services should fulfil personal user needs in varying circumstances, a bundle of services is more capable of providing the right 'bundle of benefits' (Kotler, 1999). With the advent of the open mobile internet, as opposed to the operator dominated walled garden approaches, there are more and more opportunities for mobile services and bundles.

We consider bundling as the sale of two or more separate products or services in one package for a special price (Guiltinan, 1987). In the specific case of mobile services, a bundle can be a packaged set of mobile services, that may also include a complementary mobile device, a mobile (data) subscription, or other types of services (e.g. internet services) (Teerling, Haaker, & De Vos, 2007).

Notwithstanding the wide application of bundling strategies still little is known about what constitutes a successful bundle. Much of the literature on bundling is about applying analytical methods to determine when a pure component strategy, a pure bundling strategy or a mixed bundling strategy should be followed (Stremersch & Tellis, 2002). Research on service bundling focused mainly on bundling of services with auxiliary or support services (Simons & Bouwman, 2004), and not on complementary services that are equal to the core service, i.e. voice communication services, mms-services or other mobile data-services. Moreover there are no clear design guidelines for the design of service bundles, neither for the underlying business models (Ballon, 2004). This chapter therefore focuses on CDIs for the development of (business models for) services bundles and analysis of existing bundles in the mobile domain.

We specifically address CDIs for mobile service *bundles*, i.e. additional design issues that directly relate to the *bundling* of mobile services. We consider in particular CDIs that influence the customer value of mobile service bundles' business models. The focus is on the service and technology domain, because design choices in these domains are closely related and significantly impact the customer value of a service offering.

12.1.1 Critical Design Issues for Mobile Service Bundles

The central issue in the service domain is 'value'. Value is seen as the perceived benefits and total costs (or sacrifice) of (obtaining) a product or service for customers in target markets (cf. Chap. 2). Compared to competitive offerings, a service must be considered better, and deliver the desired satisfaction more effectively and efficiently. Important design issues that influence the customer value of service bundles are discussed below. First we consider new design issues in the service domain, i.e. *Bundle Focus, Bundle Composition* and *Bundle Strategy*. This is followed by a discussion of CDIs *Pricing* and *Branding* from a bundle perspective. Finally we discuss bundle related CDIs in the technology domain, i.e. *Service Integration*, which is a new issue, and *Management of User Profiles*. Both issues may influence the customer value of a bundled offering.

- *Bundle Focus.* An important design issue is choosing a profitable target group and associated with that determining the bundle focus and scope. Should the service bundle have a clear focus (theme), or should it be broad in scope? Research on bundling reveals that most bundles are build up around one or more core services that carry the main proposition of the bundle. Some bundles are narrow in scope and focus on a niche market with very precise benefits. Other bundles focus on mass-market segments with a broader scope of (related) benefits, e.g. all-inclusive holiday packages.
- *Bundling Strategy.* Service providers can opt for three different bundling strategies: pure component strategy (unbundled offer), pure bundling strategy (components available only in bundled form) and mixed bundling strategy (components available in bundled form as well as separately) (Adams & Yellen, 1976). Bundle strategy has changed the granularity of service offerings due to the internet. For example music can be bought per song as downloads in online shops as well as per album, which provides for an example of mixed bundling. Also holiday packages are still available but users can more and more assemble their own holiday from available services (e.g. flights, hotels, car-rental, insurance) and service elements (e.g. priority boarding) that can be booked online. For mobile services often a special form of mixed bundling occurs, i.e. services are only available as add-ons to basic services or basic service bundles. We will call this add-on bundling. For example mobile data services are often add-on services to basic voice and messaging services. Pure bundling is found with Microsoft Office, i.e. a bundled offering of a word processor, spread sheet, database management and presentation tool. The actual choice for a particular strategy depends on aspects related to

the perceived customer value, e.g. degree of complementarily and integration between services (Stremersch & Tellis, 2002).

- *Bundle Composition.* A bundle can be composed of related and unrelated services. Services may be reinforcing (e.g. communication and presence information), complementary (e.g. mobile phone and subscription), unrelated (e.g. ring tone and weather information service) or competing (e.g. ring tone service A and B). Harlam, Krishna, Lehmann, and Mela (1995) suggest that bundles composed of complements have higher purchase intent than bundles of similar or unrelated products. Empirical research among users helps to get insight into which service bundles may be favored by users. Bouwman, Haaker, and De Vos (2007) have shown that add-on services can be subdivided in enhanced services and supplementary services. Enhanced services enhance the core experience of the bundle. For example a service that provides real-time traffic information could enhance existing in-car navigation by avoiding traffic congestion. Supplementary services enhance the core experience in new directions. For example audio books available via an in-car navigation system do not enhance navigation itself but intend to make travelling by car more pleasant. Bouwman, Haaker, et al. and Haaker and De Vos (2007) find that enhanced services increase bundle value more than supplementary services.

- *Pricing.* Price discrimination refers to charging different prices to different customers for the same goods or services (Shapiro & Varian, 1999). Shapiro and Varian distinguish between two forms of price discrimination, i.e. personalized pricing and group pricing. Klein and Loebbecke (2003) consider bundling as a mechanism for price discrimination. Empirical data shows that differential pricing is already widespread in industries that exhibit large fixed costs like airlines, telecom or publishing (Varian, 1996). For example in the telecom industry differential pricing may be based on customer characteristics, e.g. preference for pre-paid or post-paid, group characteristics, e.g. student discounts, product characteristics, e.g. tailored bundling of specific services and features (e.g. device, voice and data services), and volume (the more minutes you call the cheaper the price per minute). In the airline industry business travelers are typically time-sensitive and leisure travelers typically price-sensitive. To effectively differentiate between business travelers and leisure travelers the airline industry offers cheaper tickets only if a weekend is part of the trip (Klein & Loebbecke). Stremersch and Tellis (2002) distinguish between price bundling and product bundling. They speak of product bundling whenever there is some degree of integration between the bundled products or services. Price bundling refers to bundles with no integration between the services but only a price discount. Price bundling can be a valuable strategy when economies of scope or scale exists, for example in bundling information goods or services (Bakos & Brynjolfsson, 1999).

- *Branding.* Branding is also found to have a profound effect on consumers' evaluation of product or service bundles. Although a manufacturer may decide to introduce a new product on its own, he or she also may opt to promote it through bundling with an existing product which carriers either the same brand name or a different brand name. Results of Grace and O'Cass (2005) indicate that brand evidence along with advertising and promotions significantly influence consumer satisfaction, attitude, and behavioral intentions towards service brands. The findings of Simonin and Ruth (1995) suggest that a brand umbrella (within brand) bundling strategy would enhance positive consumers' evaluation of the bundle and of the new product. Moreover, consumers' reservation prices for the new product can be artificially raised for a not so-well-liked brand through a clever association with a well-liked tie in.

- *Service Integration.* The extent to which services in a bundle are integrated influences the customer value of a bundle. Due to integration the value of the bundle may be more than the sum of the value of the individual elements. For instance, Microsoft Office is a product that bundles a word processor, a spreadsheet, a database and a presentation tool. Service integration allows users for instance to copy and paste pictures made in Powerpoint into MSword and vice versa. In the mobile domain integrated search and navigation services – involving interactive maps, contact lists and yellow pages – provide for a value adding form of service bundling (De Vos, Haaker, Kleijnen, & Teerling, 2008). The question remains how specific services and bundles have to be composed. In Chap. 6 a composition method was discussed that supports the process of bundling several services to each other, in order to make the execution of a specific task, or the fulfillment of a need, possible. Service integration can be included in the composition process as additional functional requirements.

- *Management of User Profiles.* For personalization of a service, a user profile that contains user context, preferences and behavior must be created and maintained. For instance, for MSN Messenger the Instant Messaging server keeps a profile for each user. A privacy statement is issued to users about the protection of the provided data. The management of this profile, i.e. creation, use, maintenance and access to the profile, requires functionality that may be realized in different ways. Balancing is needed between user involvement and automatic profile generation, and between privacy and access to users' profiles. We introduce three alternatives for profile management.

- *Self-management* would imply that users are the initiators when it comes to manage specific issues, such as privacy, service bundling preferences, etc. This control could be done either manually by asking users to respond dynamically or through a number of (domain and user specific) preferences and rules that can govern self-management behavior.
- *Group Management.* When a group of users share a service or service bundle, control issues become more complicated. A number of factors influence this control such as the roles of the users in the group, possible hierarchical relations, the type of information to be controlled, and the context of users.
- *Autonomous Management.* Due to the huge amount of information that need to be controlled, and due to the design requirements of flexibility and ease of *use*, there will be an ongoing issue of balancing between the amount of flexibility required and the involvement of users. Therefore, a third way of controlling, autonomous control can be applied as well. Autonomous control basically implies that software will manage the applications based on predefined preferences.

Our literature review in the previous section revealed the following critical design issues for service bundling to be relevant: bundle focus (mass, niche market), bundle strategy (unbundled, mixed, add-on, pure bundled), bundle composition (core, complementary), pricing (personalized pricing, group pricing and versioning), branding (brand umbrella or not), service integration (fully, partly integrated, independent), and user profile management (self, group, and autonomous).

12.2 Revised Causal Framework Explaining Customer Value

In Chap. 3 we presented a causal model for explaining the customer value of mobile services' business models (see Fig. 3.1). Based on the literature review we extended this model to include service bundle issues, see Fig. 12.1 (Bouwman, Faber, & Haaker, 2005). This revised model explains the customer value of mobile service bundles' business models. Subsequently we look for evidence of these bundle issues in several cases of mobile service bundles.

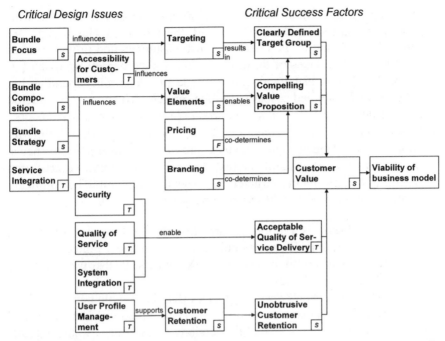

Fig. 12.1. Revised causal framework explaining the customer value of mobile service bundles

12.3 Explorative Case Studies

We analyzed thirteen illustrative cases of existing mobile service bundles (Bouwman, De Vos, et al., 2005). The cases studies were executed in various projects. Additionally some new cases were obtained through an internet search and subsequent desk research. Selection criteria for the bundles were (1) that they should comprise of at least one mobile service element, (2) that they provide illustrative examples of current service bundles and (3) that they collectively give an overview of different service bundle practices. In the cases we explicitly focused on illustrations of the CDIs related to service bundles as presented in the previous section. Mobile services are generally classified into four broad categories, i.e., information services, communication services, entertainment services and, transaction services (cf. Chap. 4). The thirteen service bundle cases were allocated to five categories, loosely following the proposed classification:

- *Operator Portals.* Most mobile operators provide their customers with a portal to access mobile services of third parties. The portals are generally restricted to the operator's own customers and provide access to all kinds of services of third parties in a walled garden approach (cf. Chap. 4). Portals usually do not provide pre-determined bundles but rather provide bundled access to mobile services. The portals considered in this study were *KPN i-mode, Vodafone Live!* and *Orange World.*
- *Information Oriented Bundles.* These bundles provide primarily information services. The services provide for information (news, weather) that is of general interest, or specific information structured around a specific theme targeted at a specific segment. The following information oriented bundles were considered:

    - *P-Info.* P-info is a mobile information service for police officers in the Netherlands. It enables police officers to query police databases and national criminal investigation registers. It also provides mail and calendar services.
    - *CNN Mobile News Services.* CNN offers browsable news, breaking news alerts via sms and mobile TV (via limited number of selected service providers) as mobile versions of existing TV and internet services.
    - *TimeSpots.* TimeSpots is a location-based mobile tourist guide. The bundle combines information and navigation services with communication services, which run on a dedicated mobile device that can be rented.

- *Community Oriented Bundles.* These bundles combine communication services with information and/or entertainment services to support communities, i.e. formation and preserving of communities. The following community oriented bundles were considered:

    - *Botfighter.* Botfighter is a location-based mobile game that lets users play with their mobile phone via SMS. On a website, the players can upgrade their robots, buy weapons, view high scores and get information on their current mission. The mobile phone acts as the player's radar and weapon.
    - *Blah!.* Blah! is a lifestyle oriented mobile community. It provides its members with downloadable content (such as ringtones) and with information channels. Members can browse through each other's profiles and interact with each other (play games, chat, etc).
    - *MSN Messenger.* MSN Messenger started out as a text based real time instant messaging and presence service based on a buddy list.

Gradually the bundle expanded to include more advanced communication and context sharing services, alert services, and information and transaction services.

- *Interaction Oriented Bundles.* These bundles primarily provide services that allow for interaction between the user and knowledgeable professionals or systems. The following interaction oriented bundles were considered:

  - *Onstar.* OnStar is a mobile service offered by General Motors for owners of selected GM automobiles. It provides information and assistance services including emergency services, as well as voice calling services. Onstar works with a dedicated car-mounted device.
  - *Predoc.* Predoc offers a disease based information and prevention program to chronic patients. The program includes personalized information, personal counseling on medication, discounts on information sessions and medical checks, and medication alerts via SMS.

- *Transaction Oriented Bundles.* These bundles are primarily provided for transactional services, for example a payment service, online shopping or bank information services. The transaction oriented bundles considered were *Postbank mobile information services* and *ABN AMRO mobile services*. Both provide a bundle of mobile banking services, i.e. transfer of monetary value, and information services like account checking and stock market SMS alert services.

12.3.1 Critical Design Issues in the Case Studies

In this section we look at how the CDIs occur in the explorative case studies. For each of the different bundle types we discuss the design choices for the bundle related CDIs.

Operator Portals. By nature the scope of the operator portals is very broad. They target a mass-market. The portal branding is such that the operator brand is always clear. However, often the content providers' brands are used as well. Access to the portal is always bundled with a basic voice subscription, or, in the case of pre-paid, separately billed. The third party services available via the portal are offered and priced completely independent, i.e. an unbundled strategy is followed by the portal and content providers. The content services are not price bundled nor product bundled (Stremersch & Tellis, 2002), i.e., bundle discounts do not exist, and content services in the portal are not integrated in any way. Customers select their services on a monthly or per use basis. The critical design choices are summarized in Table 12.1.

Table 12.1. Critical design choices for operator portals

Case	i-mode, Live!, Orange World
Composition (core)	Information, Entertainment, Transaction and Communication services
Composition (complementary)	–
Focus	Mass market
Strategy	Unbundled
Pricing	Data subscription and per service subscription
Branding	Operator brand
Service Integration	Independent services
Profile management	Self management

Information Oriented Bundles. The core of the information oriented bundles is obviously information services. In the case of P-info and TimeSpots these are complemented by communication services and, or entertainment services. Both P-info and TimeSpots have a niche focus, which provide specific information services to police officers and tourists, respectively. CNN Mobile has a mass market focus and provides generic news services. P-info and TimeSpots are examples of bundling of services with a dedicated device. Both P-info and TimeSpots follow a pure bundling strategy, i.e. all services are part of a bundled offering. On the other hand CNN Mobile has an unbundled strategy – the services are offered completely independent with heterogeneous pricing schemes. TimeSpots is offered as a daily rental, i.e. device plus services, for a flat fee. P-info is a niche business service where part of the value is clearly derived from integration between included services. Results of a search can for example be used to start another query or to begin a voice communication. For the TimeSpots bundle also integration between services exist, e.g. guided city tours with real time navigation. Profile management is not an issue for the studied information bundles. Users mainly select the information they want when they need it. The critical design choices are summarized in Table 12.2.

Table 12.2. Critical design choices for information oriented bundles

Case	P-info	CNN Mobile	TimeSpots
Composition (core)	Information services (mobile queries); navigation; task reporting	Browsable news, SMS breaking news alerts, mobile TV	*Information services*: city guide, listings; Finder service; Navigation; Audio tours
Composition (complementary)	Office applications	–	E-mail; voice; chat with other users; limited internet access; gaming; device with camera; vouchers, discounts
Focus	*Niche*: police officers	Mass	*Niche*: tourists
Strategy	Pure bundling	Unbundled	Pure bundling
Pricing	–	Heterogeneous and operator dependent	Flat fee for device and services
Branding	Police branding	Content provider branding	New brand
Service Integration	Integration	Independent	Partly integrated
Profile management	*Central management*: profile. *Autonomous management*: location	Self management	Central management; *Autonomous management*: location

Community Oriented Bundles. In the community oriented bundles the communication services form the core. The communication services are complemented with information and entertainment services, although for the Botfighter case the entertainment services are prevalent. Botfighter basically creates a gaming community with a narrow focus. The communication services enhance the gaming experience, e.g. via clan formation. MSN Messenger and Blah! are broader in scope. They support group formation and communication within user's social network via text messaging and presence information. MSN Messenger works via buddy lists under the full control of the user. The formation of buddy lists is a form of group management, as this depends on joint and reciprocal agreements between involved buddies. With Blah!, community formation or matchmaking is based on user determined profiles. The profiles contain information on personal preferences for music, going out, etc. Although some autonomous management occurs in the selection of potential matches, the final decision regarding 'matching' lies with the users. In the Botfighter case the location of game participants is determined automatically via cell-ID. Clearly profile management is an important issue for community oriented bundles.

Integration between services clearly exists in the Botfighter and MSN Messenger case. For example the robot profile in Botfighter is created on a website and subsequently used in the mobile game. With Messenger most services are integrated with the buddy list and can be evoked from the buddy list environment. For the Blah! case integration is basically absent. Blah! can be considered a 'lifestyle' portal which provides access to multiple information channels. The matchmaking and communication services between Blah!-members provide for the community aspects. For Blah! and MSN Messenger the bundling strategy can be characterized as add-on bundling. With Blah! the information channels are freely available whereas the SMS communication service is offered as a premium add-on service. Similarly, for MSN Messenger the core communication services, as well several complementary services, are freely available to all subscribers and the premium services are provide as add-ons to the core. The bundle strategy in the Botfighter case is pure bundling, i.e. all services in the bundle are available to users once registered. Pricing is not a big issue in community oriented bundles as most services are for free, except for some premium (SMS) services. The critical design choices are summarized in Table 12.3.

Interaction Oriented Bundles. Interaction oriented bundles are more targeted towards niche markets. Onstar's core proposition is about safety, whereas Predoc focuses on staying healthy. Both bundles have human assisted services as core, i.e. Onstar operators assist during emergencies via a direct voice link and pharmacies associated with Predoc provide counseling on medication. Onstar started out with a clear focus on emergency services, but extended its bundle beyond the safety theme with supplementary services like voice calling. Predoc's complementary services basically enhance the core service.

Onstar sells single services as well as bundled service packages for a special price, i.e. Onstar follows a mixed bundling strategy. Predoc on the other hand provides a pure bundled offering. The subscription based pricing model as used by Onstar is quite typical for interaction bundles, as other revenue sources like advertising or data traffic are not relevant. A subscription to Predoc is actually free for the customers of the pharmacy chain providing the service. Both Onstar and Predoc are positioned as separate brands although closely tied in with the provider's brand.

Table 12.3. Critical design choices for community oriented bundles

Case	Botfighter	Blah!	MSN messenger
Composition (core)	Robot profile (via internet), game (get mission, scan, fire) via text messages	Text based mobile communication (sms), profile matching	Text based communication and sharing presence info
Composition (complementary)	Text based mobile communication, internet chat, game information	*Entertainment*: games, quizzes, ring tones, wallpapers; *Information*: music, theatre, events	Voice and video communication; SMS service; profile matching; sharing info on music listening; price comparison; search service; emoticon trade; buying/selling via buddies; music download
Focus	*Niche*: gaming	*Mass (youth)*: wireless community	Mass
Strategy	Pure bundling	Add-on bundling	Add-on bundling
Pricing	Premium SMS	*Core*: Via SMS; *Complementary*: data traffic	*Core*: data traffic; *Complementary*: free or pay per product (emoticons)
Branding	Tie in with operator brand	New brand	Microsoft
Service integration	Integrated	Independent services	Integrated
User profile management	*Self management*: robot profile; *Autonomous management*: location	*Self management*: profile; *Group management*: matchmaking and group formation	*Self management*: preferences and context; *Group management*:buddy lists

The Onstar services are integrated in the sense that they are offered via a dedicated device and call centre. The bundle is also clearly integrated with a GM car. Profile management is not a big issue here. For Predoc the necessary medical information is already available at the pharmacy and is basically fairly static. For Onstar the car's location is determined automatically. The critical design choices are summarized in Table 12.4.

Transaction Oriented Bundles. The transaction bundles considered here focus on mobile banking. Mobile banking bundles typically combine transaction services (money transfer) with information services (account checking, alerts). Mobile banking is typically seen as an add-on to internet banking. The information services and the transfer of money are usually

Table 12.4. Critical design choices for interaction oriented bundles

Case	OnStar	Predoc
Composition (core)	Emergency Services, Air Bag Deployment Notification	Semi personalized paper newsletter; personal counseling (on medication)
Composition (complementary)	Personal calling; Information/convenience; Stolen Vehicle Tracking; Remote Door Unlock; Driving Directions	Information market; SMS news alerts on medication; health tests for reduced price
Focus	*Niche*: safety	*Niche*: health
Strategy	Mixed bundling	Pure bundling
Pricing	*Core*: subscription; *Complementary*: subscription and per use	*Core*: subscription; *Complementary*: SMS
Branding	Tie in with GM brand	New brand
Service Integration	Integrated	Independent
User profile management	*Self management*: user preferences; *Autonomous management*: collection of context information	Self management

free of charge, except for the cost of airtime. Mobile banking has a broad mass market focus with rather generic services. Bundling is basically pure in these cases as the bank's consumers have in principal access to all these services. The services are offered via different mobile channels, e.g. via SMS or regular calling, via a browser on the mobile device and/or via an operator portal. Branding is basically that of the bank brand although also a form of co-branding occurs in the operator's portal. The mobile banking services themselves are not integrated, but together they are an integrated part of the internet banking environment. Users manage their own profile via the bank. The critical design choices are summarized in Table 12.5.

Table 12.5. Bundle characteristics for the transaction oriented bundles

Case	Postbank mobile banking	ABN-Amro mobile banking
Composition (core)	Mobile banking; SMS alerts and order status	Mobile banking; SMS alerts and order status
Composition (complementary)	Revalue pre-paid calling	Financial news site
Focus	*Mass*: banking	*Mass*: banking
Strategy	Unbundled	Unbundled
Pricing	Price per minute; premium SMS	Price per minute; premium SMS
Branding	Bank brand and co-brand with operator	Bank brand
Service Integration	Independent	Independent
User profile management	Self management	Self management

12.4 Cross Case Comparison

The operator portals and the mobile banking bundles have a mass market focus. The information and community oriented bundles can be both mass market and niche market oriented, whereas the interaction oriented bundles have a niche focus. The bundles with a mass market focus seem to follow two distinct approaches. The first approach is to provide generic services that appeal to a large number of customers, e.g. bundles for mobile banking, generic news services (CNN) and presence and instant messaging services (MSN Messenger). The providers of these bundles follow an unbundled or add-on bundling approach. The second approach, followed by the operator portals, is offering a very large number of services (sometimes literally hundreds), such that there are services for every taste. The portal operators also follow an unbundled approach where pricing is service specific.

Providers of bundles with a narrow focus, e.g. P-info, TimeSpots and Predoc, aiming at niche markets with more specific services, follow mostly a pure bundling strategy. Only Onstar is found to pursue a clear mixed bundling strategy.

Integration between services is most often found for bundles with a niche focus. Providers of these bundles use integration to create added value, e.g. in the case of P-info, Botfighter, Onstar and TimeSpots. Bundling with a dedicated device is found for niche bundles, i.e. P-info, Onstar and TimeSpots. Bundles targeted at mass markets typically make use of the user's personal mobile device.

Management of user profiles is most important for community oriented bundles, much less for information and interaction oriented bundles. Control over user's profiles is typically based on self management, while relationships with group or community members are based on mutual agreements, i.e. group management. For P-info, a business service, profiles are maintained at a central location. Location information in services usually automatically collected.

12.5 Conclusion

This chapter presents additions to the STOF CDIs with respect to service bundling. Whereas the focus in Chap. 3 was on CDIs for mobile services' business models, this chapter particularly addresses new CDIs that arise when mobile service *bundles* are considered.

We can conclude from our exploration of mobile service bundles that bundles normally contain two main types of service elements: core services and complementary services. What is core and complementary depends on the intended bundle value and the scope of the bundle.

The bundle related design issues show that bundles vary in bundle focus, strategy and composition. Some combinations of service elements can be observed more often than others. Typically complementary services, which strengthen each other, are bundled together, especially in bundles with a niche focus. *Pricing* schemes (and therefore the implied revenue models) differ from case to case. Another driver for bundling is *Branding*. Brand names of vested players may make service bundles more attractive for customers. Our technical design issues show that system integration can be an important driver for bundling and creating value as well. Finally, management of user profiles is managed in different ways but user control is an important driver there.

Starting from the STOF model, in particular the CDI, we identified the CDIs for the business models of mobile service bundles that relate to the creation of customer value. The critical issues and causal relationships that we found in Chap. 3 remain valid but we have identified a number of additional bundle related issues that need to be taken into account when mobile service bundles are considered, e.g. bundle composition. Also some existing CDIs, i.e., *Pricing* and *Branding*, require a bundle view to adequately deal with them in a bundle context.

Although not addressed in this chapter, naturally also additional bundle related CDIs for creating network value may be identified, e.g. relating to economies of scale and scope. The work in this chapter should be considered as a starting point for further conceptualization of issues in the area of service bundling and service compositioning (cf. Chap. 6), since the knowledge base of 3G+ bundles is limited. Future effort may for example focus on issues for advanced mobile offerings, e.g., personalized context-aware mobile service bundles.

Chapter 13 Designing Mobile Remittance Services in Developing Countries

H. Bouwman and J.-C. Sandy

In this chapter, we illustrate how the STOF model can be applied to analyze business models, using an SMS-based service that is deployed in a developing country, i.e. Smart Padala in the Philippines. In addition, we use the STOF method to design a business model for the introduction of a remittance service in Haiti. Remittances are understood as transfers of money from one person to another, in practice most transfers are by foreign workers to their home countries.

13.1 Mobile Services in Latin America

The exponential growth of mobile communication in developing countries provides the potential for a wide range of services. Although some may believe that mobile services remain the preserve of the relatively wealthy, people living in developing countries, despite the fact that they have little income, use mobile services mainly in the form of prepaid subscription and through handset sharing. In developing countries, in contrast to Western industrialized countries, wireless telephony and applications are not used as complementary functionalities, but rather as a substitute for traditional telephony.

To provide an indication of recent developments in the mobile market of developing countries, in 1996 Latin America had six million mobile subscribers, a number that increased to 118 million by 2003, and to 171 million by 2005 (GSMA/IFC). Eighty percent of mobile services are used on a prepaid basis. The average revenue per user (ARPU) is less than US\$ 20 per month. Operators offer low-cost handsets with limited functionalities to lower-income subscribers (Rojas, 2005), providing basic mobile telephony, while more affluent customer segments have access to data-enabled value-added services based on GPRS/EDGE.

Although Latin America appears to be the fastest growing region in the world with regard to the Internet, there are several obstacles to its development. Due to the relatively low income levels, many people have no access to a telephone, and even fewer people can afford a computer. Computer literacy is relatively low and Internet access is still comparatively expensive. Most users access the Internet via public terminals or at work (*ECLAC*, 2003). Compared to Western Industrialized societies, Internet use lags behind by several years. As a consequence, the development of e-commerce in developing countries has been slow, in part because of low credit card penetration and a poor infrastructure.

It is expected that developing countries will bypass traditional e-commerce models and that they are leading when it comes to the adoption of mobile services. Obviously, there already are developing countries that have successfully introduced mobile commerce, for example in the form of mobile remittance. In this chapter, we focus on the design of business models for mobile remittance services in developing countries. Although the conditions discussed in Sandy and Bouwman (2006), which include regulation, technological innovations and market-related issues, are important to the success of mobile services, they lie beyond the scope of this chapter. As a starting point for understanding the relevant issues of developing countries, we analyze an example of a successful mobile remittance service called Smart Padala, which has been deployed in the Philippines since 2004.

13.2 Smart Padala

The aim of our analysis of the Smart Padala business model, which is based on reports and Internet sources (GSMA/IFC, World Resources Institute, Smart), is to identify the issues that are characteristic of mobile services in developing countries.

13.2.1 Service Domain

Smart Padala is an SMS-service that has been provided since 2004 by Smart Communications, the Philippines' leading wireless services provider providing a range of m-commerce services. Globe Telecom, the country's second largest mobile operator, offers a similar service known as G-Cash, which was also launched in 2004.

Smart Padala is an international text-based cash remittance facility. It is estimated that Filipinos working abroad send each year between \$14 billion and \$21 billion back home. The service enables recipients to be informed immediately when, for example, someone in the United States has wired up to a maximum of Pesos 10,000 to the recipient's Smart Money number that is linked to the recipient's Smart cell phone number. The recipient can then collect the money from one of the numerous distribution cash-in-centers, including banks and ATMs, using their Smart Money card. Once the cash has been transferred to the Smart Money account, it can be used in shops and restaurants. The cash value may also be used to load airtime, pay utility bills or transfer money from the Smart Money card to another card. In addition, Smart Money has served as a platform for other m-commerce services.

Smart Communications' mobile financial services are used by 2.5 million people, many of whom use no other formal financial services. In addition to Smart Padala, Smart offers its customers a set of other financial services provided by its main product SMART Money, including (1) cash deposits and withdrawals, (2) bill payment and direct deposits from payroll, (3) Smart Load: credit recharging in small amounts, (4) Smart PasaLoad: transfer of airtime credit from one user to another, and (5) cashless purchasing at shops where retailers either have a Smart Money account or are MasterCard-enabled.

13.2.2 Technology Domain

As is the case with other SMS-based value added services, the infrastructure, hardware and software needed to deploy the remittances services are add-ons to the network. For financial reasons, a light-weight solution has been chosen. The most important aspect of the technology is the implementation of a security scheme operating under the existing SMS set-up and for which no additional hardware or software is needed, either in the SMART network or on the part of the customer. The service is compatible with GSM 'Phase 2 Plus' SIM standards and uses SIM-based memory and menu capabilities. A simplified network architecture is presented in Fig. 13.1.

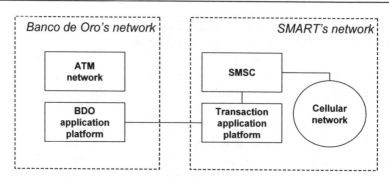

Fig. 13.1. SMART architecture

13.2.3 Organization Domain

The main providers of SMART Padala are Smart and its financial partner Banco de Oro (BDO). The roles of the two main partners are delimited very clearly: SMART serves as a transport system and hosts customers who have no relationship with the bank, and it offers customers the opportunity to send text messages to the bank's systems. Banco de Oro is the retail bank processing the mobile financial transactions. The service operates entirely within the limits and the jurisdiction of the central Bank, and as such Banco de Oro takes responsibility for auditing, fraud management, account security, etcetera. The clearly defined roles of the two major partners provide an enabling regulatory environment: the telecommunication regulatory body has no interest in regulating this service, since it does not involve unusual telecommunication aspects, and as far as the financial regulatory body is concerned, the service facilitates regular financial transactions. A third important group of partners is made up of the associated retailers located overseas and in the Philippines.

13.2.4 Finance Domain

With minor exceptions, SMART receives its income entirely from the SMS charge levied for each transaction: $0.05 per user-generated SMS activity; $0.02 for retail purchases; $0.06 for ATM withdrawals though Banco de Oro and $0.21 for non-Banco de Oro ATM withdrawals; 1% fee for cash deposits and withdrawals in the retailer network. These charges are attuned to the low transaction fees charged in the Filipino mobile market, normally $0.02 for standard SMS text messaging and $0.60 for mobile credit top-ups. One of the positive outcomes reported by Smart is mobile wallet revenue growth of 46% in 2005, and the use of Smart's remittance service by one million overseas Filipino's, sending $50 million per month in international transfers. As a result, the Banco de Oro has a cash float of $10 million, over which interest is generated. As far as the telecom operator is concerned, one of the positive intangible outcomes is a significant reduction in customer churn.

13.3 Designing the Haiti Case

Based on the insights provided by the successful Smart Padala service, in this section we focus on the design of a similar service in another part of the world, namely Haiti. To this end, we use a combination of the STOF method's quick scan and action research, which means that information that is relevant to the design of the business model is exchanged with the actual service designers.

13.3.1 The Haitian Context

Many Haitians live overseas and each year send up to $1.65 billion back home, representing about a third of the country's Gross National Product. About 1.1 million adults in Haiti receive such remittances; typically 10 times a year and on average $150 per remittance. What is particularly relevant here is the fact that 83% of the money is sent through remittance companies, 6% through banks, 4% by mail and 4% through friends or relatives traveling to Haiti (Boyd & Jacob, 2007). The social and economic relevance may be clear, and the opportunities available to new players in the remittance business and the introduction of new paradigms, such as mobile remittances, is obvious.

13.3.2 Service Design

The service is similar to that of Smart Padala, i.e. SMS-based mobile remittance. Because the potential customers of this service have low incomes, prepaid solutions and low fees are a logical choice. There are two key products: the ability to load prepaid airtime credits over the air, and the transfer of cash and airtime credits between customers. The fees set by the operator for such prepaid top-ups or credit transfers are low. Typical top-ups of US $0.57 to US $47 were allowed by the networks, while transfers between customers of both cash and airtime credits of as little as US$ 0.04 were permitted. Because typical customers are familiar with

SMS, it is easy for subscribers to start using remittance services and other mobile banking services.

13.3.3 Technology Architecture

The service depends on the use of security technology and on the basic SMS or Unstructured Supplementary Service Data (USSD) technology. The recipient of the remittance needs a 16 digits 'Smart Money number', which can either be retrieved by sending an SMS or by utilizing a SIM toolkit menu. This 'Smart Money number' is stored in the SIM and is accompanied by a password the recipient has to enter with each transaction. Before the system can be deployed, the security and encryption features have to be developed. Moreover, the technology can have an effect on the ease of use of the SMS service. Illiteracy rates in Haiti are high, which means that a simple system has to be designed that can be provided by menus developed through interactive USSD or, preferably, SIM Toolkit menus.

13.3.4 Organizational Arrangements

The service is highly dependent on the involvement of key actors such as distribution points and banks. As Faber and Bouwman (2003) point out, their participation is crucially important to the success of any mobile payment service. This means that it is of the utmost importance to include a bank that has direct ties to the Mobile Telecom Operator offering the service, to delineate revenue lines. Alternatively, operators can start carrying out encashment activities themselves.

The following roles are distinguished in providing the remittance service:

- The sender, the person who sends the money, goes to a Foreign Remittance Center (FRC) and provides the recipient's phone number, electronic money number and a valid ID.
- Foreign Remittance Partner or Center (FRC), collecting the money to be transferred. The FRC sends an SMS to the Recipient's phone number.
- The mobile network operator providing the software and hardware required for the electronic transfer of the money. The sender should receive the confirmation message via the cell phone from the remittance center.
- The recipient receives an SMS message (Beneficiary receives cash) that 'cash' has been sent to his or her cell phone.

- The encashment partner responsible for transferring the electronic money into cash. The recipient goes to an encashment center and cashes the money.

13.3.5 Finance Domain

The revenues need to feed all the partners within the value web. To begin with, the remittance center – the location where the money is deposited – will most likely receive a percentage. Secondly, the mobile operator will benefit from the increase in SMS traffic. The operator deploying this service will reinforce the relationship with its customers, which will reduce the churn rate. Customers are less willing to switch to other providers once their telephone number has been linked to an 'electronic money number' or identifier. Marketing the service is a costly affair. Getting people used to the service either implies launching promotions which are free, or creating some sort of viral marketing, as well as educating the subscribers with regard to the use of the service. Other investments are related to technology, i.e. the deployment of SIM toolkit and the deployment SIMs of at least 64K to run the mobile commerce applications. The bank's incentive is related to the opportunity of reaching people who do not normally use its services. If enough money is captured from remittances, the float and interest provide an additional benefit to the financial institutions.

13.4 Conclusions

As far as the mobile remittance case is concerned, the STOF model and method proved helpful in analyzing and discussing design issues in a systematic way, and in addressing critical design issues that are relevant during the service design phase. Based on the analysis, a more detailed business case can be constructed on the basis of the holistic framework for the design of mobile remittance services. The analysis of the SMART case in the Philippines helped us identify the success factors and design issues that would make it easier to deploy a similar service in Haiti. At a service level, user-friendliness is essential for an illiterate target group, and the services for low income users need to be characterized by low fee/high volume. At an organizational level, involving a bank is crucially important. It is clear that one of the key success factors is the presence of a partnership with a bank similar to the one with Banco de Oro in the case of Smart. At a technological level, low-tech alternatives – such as SMS,

USSD and SIM toolkit menus – are relevant. The widespread use of and familiarity with SMS technology is another factor. Problems regarding identification and user-friendliness can be overcome by introducing either USSD or SIM toolkit type of menus. At a financial level, both tangible and intangible benefits need to be taken into account. We expect that a dual market will develop in many developing countries, in which high-end technological services are adopted by wealthy urban professionals on the one hand, and low tech alternatives are made available to the less affluent. Based on the examples in the Philippines and South Africa, we conclude that mobile remittances and other types of mobile commerce are more likely to take off in developing countries. Moreover, due to the multiplier effect, the role of mobile services in social development and the use of mobile financial identities provide a value-added essence to Mobile commerce. Mobile commerce services will provide the tools that are currently not available to poor people, due to the costs of the technologies involved as well as the limited availability of the Internet. Mobile services will enable mobile commerce to outperform e-commerce.

Chapter 14 Assessing the Business Potential for New Mobile Services from Mock-Up Evaluation

T. Haaker and B. Kijl

Mock-ups are often developed in technology-oriented research projects to visualize the potential functionality and value of the technology under investigation. A mock-up typically visualizes a service's interface and outcome, which can be used for the purpose of demonstration. Mock-ups can be used to extract user requirements for new services at an early stage. In this chapter we illustrate how the STOF model and method can be used to assess business potential of new services from mock-up evaluation. The design of business models and assessment of business potential of mock-ups may serve different purposes. First, the business potential can be used in the selection of promising mock-ups and subsequent prototypes. It also reveals which mock-ups should not be pursued. Second, critical design issues for the business model may be identified and resulting business requirements can be taken into consideration already at an early stage. Finally, the mock-up and the initial business model can be used to interest and involve potential stakeholders. Just as the mock-up can be used to identify user requirements, the business model can be used to identify business requirements, for instance with regard to the required business roles and their interfaces. It is assumed that early identification of business requirements leads to a higher possibility of a viable business model. The STOF model and method are, therefore, instrumental in realizing a higher chance of valorizing research mock-ups and prototypes.

14.1 MobiLife Application Mock-Ups

The aim of the MobiLife Integrated Project in IST-FP6 was "to bring advances in mobile applications and services within the reach of users in their everyday life by innovating and deploying new applications and services based on the evolving capabilities of the 3G systems and beyond" (Klemettinen, 2007, p. xi; *MobiLife*, 2008). The project recognizes that the challenge of enabling large-scale ubiquitous services and applications remains. The project's primary focus was on developing technical enablers for context- and, group-awareness, and multimodality in mobile services. However, non-technical challenges regarding a user-centric approach and business viability were also addressed within the project. To that end a number of reference applications were developed by the technical work packages, to highlight the possibilities of the innovative technologies under consideration (Klemettinen, 2007). In particular, small-scale user trials and business analysis of mock-up applications were part of the project. In this context, a mock-up is a visualization of a service's interface that can be used for the purpose of demonstration, to test a design, et cetera Mock-ups are only designed to look like the real system, and do not have the functionalities that are found in prototypes.

14.2 Assessing Business Potential

In the MobiLife project the business potential of eleven mock-up applications was assessed. The assessment guided the choice of applications for which (partially) functioning prototypes would be built. The main selection criteria from a business point of view were intended customer value and market potential, in addition to more practical implementation considerations. The viability of the mock-up applications and the underlying technologies were assessed using the STOF model and method. For each mock-up application, a business model design session was organized using the approach outlined in Chap. 5. In the sessions, special templates were used in which business model components were filled in and issues that still needed to be resolved were collected. The sessions focused on components within the business model domains that were considered the most relevant in light of a session's purpose and the – limited – timeslot available.

In the next section we begin by focusing on the set-up of the business model design sessions. We provide observations regarding the process of the sessions and how it relates to the sessions' results. As an illustration we present the results for one of the mock-up applications that were obtained in the session, and their further elaboration. Finally, we address the way the results of the sessions influenced the choices in favor of particular applications to be developed in later stages of the project. We end with some conclusions regarding the use of the STOF model and method.

14.3 Mock-Up Evaluations Based on the STOF Method

In this section, we begin by describing the mock-up evaluation process, i.e. the business model design sessions. Next, we look specifically at the session's results for one of the mock-ups, the 'Context-Aware Interpersonal Communicator'.

14.3.1 Set-Up of the Business Model Design Sessions

The business model task group in MobiLife prepared the set-up of the business model design sessions. The sessions' participants consisted of MobiLife project members from various companies and organizations. Although most of them had a technical background, there were also some participants with a social science or business background. In each session, around 15–20 people took part in a brainstorm and discussion about several business model elements, focusing on potential user groups, revenue sources and value network structures.

The sessions followed a pre-defined format that was prepared by the task group. Task group members also acted as sessions' facilitators. Templates were prepared for the participants to write down their ideas about the design of business model components.

After attending a brief introduction of the mock-up, the workshop participants divided into groups of two, and were asked by the session facilitator to think for about 5 min about potential *Target Groups* for the mock-up introduced earlier. More specifically, they were asked to think about potential customers (people who may pay for the service), potential end-users (people who may actually use the service) and potential user contexts. Although in some cases end-users and paying customers can be the same people, this is not always the case. Each couple discussed potential target groups and wrote their main ideas down on the template, after which the main ideas where discussed in a plenary discussion of about 10 min. Every couple had the opportunity to present their ideas. The ideas were directly processed and summarized by a second session facilitator – using a laptop and a beamer. The results were directly visible to all session participants, which stimulated discussion and helped generate ideas.

The participants were then asked to describe *Value Drivers*: potential benefits of the service to end-users. They had to think about substitutes or alternative solutions and were asked to think about how the mock-up could be used to distinguish the product from these substitutes or alternative solutions. To answer this somewhat more complex question, the session participants (in groups of two) again had about 10 min, after which the results were debated in a plenary discussion that lasted about 20 min. Once more, all the ideas were directly summarized and made visible.

The final question was aimed at *Value Networks* by focusing on the business roles that had to be filled in order to offer the mock-up as a service in a commercial setting. The workshop participants were asked to identify the various *Business Roles* and their added value. Secondly, they had to identify potential *Revenue Sources*. Once more, they were given about 10 min to come up with ideas, after which the results were discussed in a plenary discussion of about 20 min. However, this time the couples were combined into groups of about six people and asked to develop value networks as graphical representations of business roles, and their relationships.

Evaluation of the Session's Process

The session's set-up proved very useful for the both participants and facilitators, as well as from a content point of view. By giving people the opportunity to first develop and discuss their ideas in couples and afterwards giving every couple the opportunity to communicate their findings, all the participants had the chance to share their ideas, which helped avoid a frequently occurring situation whereby only the most extrovert people share their ideas, preventing those with a less sanguine disposition from sharing their ideas, which in principle are equally valuable. Also, the idea of using a second workshop facilitator to process and share the workshop results – via a laptop and a beamer – seemed to stimulate the idea generation process, i.e. by supporting participants to build on each others ideas. The fact that people had to answer the final question by drawing a *Value Network* in bigger groups of around six people added an extra social element to the workshop. Although people liked this way of working, it turned out most of them needed more time to answer the question than was originally planned. This is partly due to an increased need for communication, but it may also have to do with the relative complexity of the question.

By alternating between small groups, bigger groups and plenary discussions where every group felt a kind of social pressure to present good ideas, everyone actively participated in the workshop. As a result, the quality and quantity of the content yielded by the workshop was high – especially when we take the limited time that was available for the workshop into account.

14.3.2 Sketch of Sessions' Results

In this section, we look at some of the results of the business model design sessions by focusing on the outcome of one session involving a specific mock-up application, called the 'Context-Aware Interpersonal Communicator' (based on Killström et al., 2006). After providing a brief outline of this mock-up, we take a look at the findings for each of the business model elements mentioned before. We conclude with a short discussion about the critical open questions that the workshop yielded with regard to the viability of the business model for the mock-up.

Description of the 'Context-Aware Interpersonal Communicator'

The Context-Aware Interpersonal Communicator (see Fig. 14.1 for a screenshot) is designed to support light-hearted, effortless interpersonal communication and helps people stay in touch and maintain peripheral awareness of each other's whereabouts and activities over extended periods of time. More concretely stated, the Context-Aware Interpersonal Communicator makes it easy for people to get information about e.g. the mood (Alice is happy), activities (George is listening to Radio 1) or location (Barbara is in Oulu, Finland) of the people in their friends' list. The service concept capitalizes on trends like being 'always in touch with your friends' and 'continuous communication between friends', and

Fig. 14.1. Screenshot Context-Aware Interpersonal Communicator

technical trends like the emergence of presence-awareness technologies. In addition, it shows the importance of asynchronous (IP-based) digital communication, and the service may actually also trigger synchronous communication traffic (where presence information like 'Jeff is very angry' can be seen as a form of conversational content triggering e.g. a voice call from a curious friend).

Results for the Service Domain

As we mentioned above, a distinction has been drawn between customers (the people who pay for a service) and end-users (the people who actually use the service). With this notion in mind, the participants identified the following customers and end-users.

Target Group. First customers could be companies that would like to use the service for marketing purposes (e.g., branding) by offering the service for free to end-users. Next associations and communities, i.e. organizations representing groups of people with common interests, could be customers. Also operators and end-users as paying customers were considered possible. Operators may offer the service to end-users for free, because the service may, as a form of conversational content, lead to increasing synchronous communication traffic. As end-users individual people, especially youngsters, were considered. More in particular people with common interests that are member of organizations that could act as paying customer for the service. Also less formal groups could become users, e.g. partners or families, or groups of friends and/or colleagues. The participants concluded that on the basis of the current value proposition, youngsters seem to be the most interesting target group. Having said that, the service may also be valuable in a business context because it can make communication more effective and/or efficient (with a more business-like user interface and value proposition). The diversity in target groups shows that group technology has value in different contexts and therefore different business models are needed for services based on essentially the same technology.

Value Drivers. A diverse set of value drivers has been identified for the Context-Aware Interpersonal Communicator. Helping people getting in touch with each other and having fun is an important driver. The service also enables communication without disturbing people and makes it easier to share one's emotional status or feelings to (a group of) people. The service could also be used to share other information/media or to plan activities or social events. The feeling of belonging to a group and having peripheral awareness of buddies were identified as specifically important value drivers.

According to the participants, the service could be seen as an addition to synchronous communication channels like instant messaging, SMS or voice communication. Unlike these channels, the Context-Aware Interpersonal Communicator should aim specifically at sharing one's emotional status and asynchronous communication.

Results for the Organization and Finance Domain

Value Network. The following business roles were identified by the workshop participants as necessary for offering the service commercially, see Table 14.1.

Table 14.1. Overview of identified business roles for Interpersonal Communicator

Business role	Activity
End user	Uses the service, creates content
Customer	Buys or commissions the service
Service provider	Delivers the service
Service developer/application provider	Develops services and provides applications
Device manufacturer	Supplies devices, terminals, etc.
Context and content providers	Provides required context information or content
Connectivity provider	Provides connectivity
Advertiser	Provides advertorial content

Note that the end-user in this service may play an active role as co-creator of content, or as prosumer, e.g. in developing new types of emotional information icons. Based on this identification, several value networks were designed, one of which is shown in Fig. 14.2. Depending on the type of customer, this model may be a 'b2b2c' model, if the customer is a business client, or a 'b2c' model, if the customer is also the end-user. In this network, the connectivity provider is an operator who may also contribute billing capabilities. Studying alternative value network structures and business configurations is helpful in finding a focal actor that may lead the development of the innovation into a viable offering and business model.

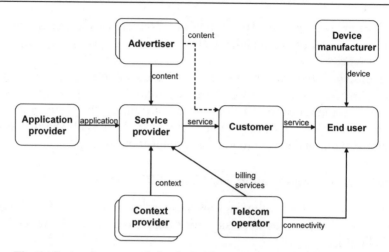

Fig. 14.2. A value network for the Interpersonal Communicator service

During the workshop, participants asked some interesting questions about the sustainability of services like the Context-Aware Interpersonal Communicator; e.g. 'Do people really want to be connected anyplace and anytime?' or 'Are people willing to share their private information via the Context-Aware Interpersonal Communicator?' According to subsequent user research (Kurvinen, 2006) the answers to these questions depend on the person with which the information is to be shared, on the reason for sharing and on the degree of user control. With regard to revenue sources, participants were skeptical about end-users willingness to pay for a service like this.

14.3.3 Business Requirements and Critical Design Issues

The results of the business model design sessions were processed further and harmonized by the business model task group in mock-up business model reports, which in turn were used to detect general issues that could have a critical impact on the acceptance and viability of service offerings based on the technology underlying the mock-ups. The generic issues point to potential bottlenecks in the future viability of the services and require special attention, preferably in the R&D phase. Challenges with regard to MobiLife type of context-aware services were further studied in a series of expert interviews (Haaker, Galli, et al., 2006).

Logical Actor on the Market. If there are clearly identifiable actors on the market, the business model is easier to realize; this usually means evolution of existing services by using new technology. If such actors do not exist or it is unclear who will take up new roles, it diminishes the chance of defining a viable business model. Based on this finding, a number of generic business models were defined and analyzed in which the (new) roles, e.g., context or trust providers, were alternatively played by different actors (Killstrom et al., 2007).

Revenue Sources. Research within and outside the MobiLife project suggests that users are not willing to pay for these kind of mobile services (Haaker, Galli, et al., 2006). This obviously decreases the possibility of new business models, unless other new revenue sources can be found, e.g., advertisers. Most MobiLife-enabled services were found to be suitable for this approach, especially since advertising could be personalized and contextualized. Based on this outcome, advertising and subsidy-based business models were analyzed further (Immonen et al., 2006).

Technology Specifics. We obtained specific business requirements about the focal technological topics of Mobilife, i.e., personalization, group-awareness and context-awareness. These requirements have an impact on the further development of the technology and its applications. For example, for personalization to be viable from a business point of view, the technology has to be sufficiently generic, such that it can be used in larger set of services (as opposed to being used only for a single service). For group-awareness, the trust and privacy solutions in services increase willingness of end-users to utilize the group-aware enabled services. Users would, for example, require easy-to-use control regarding with whom, when and for what purpose they share information about their location, as well as other information. Subsequent mock-up evaluation by end-users confirmed that users appreciate the potentially improved convenience, but they do not wish to lose control of their own schedule and habits or automating everything. Context-awareness business viability requires a continuous flow of context-related information, which may be accompanied by rising costs and organizational issues.

14.3.4 From Mock-Ups to Prototypes

The mock-up evaluations were used together with other project criteria to determine which mock-ups should be further developed into a (partially) functioning prototype. The business assessment followed a scoring approach in which researchers and task group members rated each mock-up on a number of items that link to Critical Success Factors (see Chap. 3), i.e., *Clearly Defined Target Group* (end-users and customers), *Compelling Value Proposition, Acceptable Division of Roles* and *Acceptable Profitability*. Each item was scored on a five-point Likert scale. The final mock-up rating was summarized in a rating for assumed customer value and one for assumed network value.

Table 14.2 shows an overview of all considered mock-ups. Additionally, for each mock-up the market potential in terms of market size was addressed. Typically the more specialized applications target a smaller or niche market segment. Services like FamilyMaps and Mobicar target niche markets, whereas a service like the Context-Aware Interpersonal Communicator is more generic in nature and targets a larger segment, i.e. youth in general.

Table 14.2. Overview and description of initial MobiLife mock-ups

Number	Name	Description
1	FamilyMaps	Provides location-sensitive support for families with small children when on the move
2	Infotainer	Displays news to the user in a context-aware and multimodal way
3	Wellness aware gaming	Takes advantage of the personalization improvements to provide tailored training plans
4	Emergency preparedness	Supports people taking care of relatives, friends, and/or patients in the case of an emergency
5	Mobicar	Provides a group- and context-aware service for car-sharing allowing for profile based matchmaking between drivers and passengers
6	MyLifeViewer	Provides an all-embracing view of the different aspects and activities of users' life and context-based reminder services and communication support within groups
7	TimeGems	Suggests to users potential activities within related groups, taking into account the availability and preferences of group members
8	Bus Stop	Provides real-time personalized and context-aware travel information and suggestions about relevant services to people waiting for public transport
9	Context Augmented Scheduler and Reminder	Allows end-users, groups and organizations to manage task reminders using context-aware capabilities to improve task performance
10	Context-Aware Interpersonal Communicator	Makes it easy for an end-user to automatically record, store, use and share context information
11	Tourist Info System	Supports the end-user in going around and enjoying at best a touristic visit

The main goal of the business potential assessment is not to come up with absolute numbers for each mock-up, but rather with an assessment of the relative potential of each mock-up in terms of clarity of assumed customer value, assumed network value and market size. Figure 14.3 shows the relative position of all mock-ups in a single graph. On the horizontal axis the clarity of the network value is shown and on the vertical axis the clarity with regard to assumed customer value. The size of the circle representing the mock-up symbolizes the assumed market size.

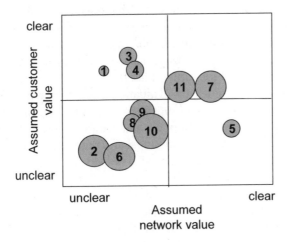

Fig. 14.3. Graphical view of business potential of initial MobiLife mock-ups

Eventually a number of eight mock-ups were pursued. As mentioned in (Klemettinen, 2007, p. 229) "primarily the applications were included neither to show breakthrough concept innovations nor user interfaces polished to the level of commercial solutions, but to show implementations that are actually following the architectural and technological frameworks". However, the reduction from eleven to eight mock-ups partly followed a business reasoning based on the mock-up assessments. For example the Bus Stop mock-up (no. 8) was stopped because it lacked a sizable target group as well as a clear customer value and network value. Instead the narrow context of use was broadened in a new application labeled Proactive Service Portal, which provides contextual reasoning to pro-actively offer services to the mobile user. The Tourist Info System (11) was removed as it lacked innovativeness given that similar services were already entering the market. The market for Emergency Preparedness (4) was considered too small and the provider value was unclear, especially regarding the division of roles, i.e. who should be the focal actor. Some elements of mock-ups were combined with other mock-ups as the separate mock-ups were not considered viable. For example parts of MyLifeviewer (6) were integrated with the Context-Aware Interpersonal Communicator (10) to form a new application called Context-Watcher (Koolwaaij et al., 2006).

14.4 Conclusion

The STOF model divides business model reasoning into four easy-to-comprehend domains. The STOF method helps to focus the design. Given the limited amount of time available, many valuable results could be obtained in a business model design session with project members. This also proved a sound basis for further analysis.

The outcomes of the mock-up evaluations, together with other criteria, were used to decide which mock-ups were continued or combined. Cross-case analysis revealed revenue models and pricing as critical design issues, which lead to research into advertising based models. The results of the sessions with regard to critical issues also helped formulating the questions to consider in the user evaluations.

In technology oriented projects, valorization is a general problem. It is often hard to bring technical research to viable offerings, for various reasons. First, the results are still far from the market, which makes it difficult to bridge the gap from R&D to marketing (Chesbrough, Vanhaverbeke, & West, 2006). This means that evaluating mock-ups from the point of view of users and businesses is useful in identifying the potential target groups and potential stakeholders at an early stage. The STOF model and method make it easier to assess opportunities provided by technological innovations in marketing terms, i.e., in terms of value proposition, target groups, competitive edge, value network, revenue model, et cetera. As such the STOF model provides a language for the valorization of technology, with in the end a greater chance that technology is turned into value creating offerings.

Chapter 15 A Standalone Digital Music Vending Service

H. Bouwman, M. De Reuver, and H. Schipper

In this chapter, we apply the STOF model and method to a case that involves a digital music vending service, called 'MusicBox', in a newly industrialized country. The service uses store locations where users can obtain digital music from digital music vending machines. The service enables users to make use of mobile devices like MP3 players, telephones and the like to listen to popular music. We used the STOF model as a prescriptive framework. Accordingly, the present chapter addresses the following question: Can the STOF model and method be used as a framework for designing a digital music vending service and underlying business model? The analysis is based on the first two phases of the dynamic business model, i.e. the conceptualization of the first ideas, and the initial market trials.

15.1 MusicBox

As a result of numerous technological innovations, the global music industry has changed considerably over the past decade. While the sales of music based on physical media, i.e. CD and cassette, have fallen, the consumption of digital music in the markets of developed countries has increased. The key factors with regard to this increase include affordable broadband access via Internet, an abundance of ripped digital music, a sufficiently high PC penetration, and low prices of portable music players. However, in the country under study – a newly industrialized country – the density of PCs is low and the penetration of broadband facilities is almost zero.

In an attempt to overcome the absence of these two key factors, a specific company is developing MusicBox, the main goal of which is to offer digital music to consumers by setting up digital music vending units (kiosks), bypassing traditional PCs and the need for broadband Internet access. Consequently, MusicBox provides the 'on-the-go' experience in a newly industrialized country. As a result, its value proposition is to enable end-users to purchase and download high quality original digital content legally, via a network of multimedia vending units, placed at convenient locations throughout the country to enable the music to be played on portable devices.

In the next section we discuss the intricacies of the music industry in the specific country under study. The subsequent section contains an overview of existing external drivers, as well as preconditions characterizing the indigenous market. After that, we elaborate on the manner of applying the STOF method in the design process. We discuss the Critical Design Issues (CDIs) involved in the MusicBox case. Next, we illustrate the various interdependencies between the CDIs of the various domains. In the final section, we present our conclusions and discuss the issues we addressed.

15.2 The Indigenous Music Industry

We begin by identifying the external technology, market, and regulatory drivers that influence the business model. The introduction of digital music and digital data-compressing techniques in general are important technical drivers, because they have a significant impact on the global music industry, in particular with regard to the distribution channels, as digital (non-physical) formats become available. The distribution channels of music are changing and new distribution channels evolve as the availability of broadband Internet and third generation (3G) mobile services enable licensed digital distribution services. Obviously, the same drivers have enabled a large-scale unlicensed p2p exchange of digital content, which thus far has proven much more important than its legitimate counterpart. However, in the market under investigation, PCs and broadband Internet are hardly available and adoption levels are low. In addition to a channel used to obtain content, a device to play digital music is also required. The development of such devices, at lower prices for devices with ever larger storage capacity, is another technical driver. In addition, the development of these devices is connected to the available digital audio formats.

Market drivers have to do with issues such as disposable income, demographics, available leisure time, and demand for leisure (Vogel, 2004). The fact that consumers in newly industrialized countries have relatively small disposable incomes implies that markets will tend to favor low prices and high volumes. However, as the economy in the countries in question is growing, so is the average disposable income of consumers. The number of digital music playing devices, like MP3 players and mobile phones with built in MP3 functionality, is growing. A more specific trend involves the music sold by the indigenous music industry, which for the greater part consists of local film music.

Piracy also affects the music industry. The absence of regulation may be considered an external driver. Relevant legislation ('Copyrights Act') is outdated and there is no suitable regulation to prevent piracy. In addition, the government is hardly capable of enforcing anti-piracy laws. Accordingly, lengthy legal and arbitration processes are not helpful in the fight against piracy. These external drivers influence the requirements and assumptions that have to be addressed with regard to the four core domains, i.e. the service, technology, organization and finance domains, of the business model.

15.3 Application of the STOF Method

The STOF model and method guide the stages of definition of requirements and assumptions, as well as determining structural preferences. The design method explicitly addresses questions regarding the four domains and takes into account Critical Design Issues (CDIs) as well as Critical Success Factors (CSFs) related to creating customer and network value (Chaps. 3 and 5). To provide answers, one can cluster the questions based on potential data sources, such as consumers, the own organization or specific partners. Such clustering is akin to defining the kind of requirements or assumptions with which certain questions deal, i.e. functional, user-related or context-related. After that, one needs to formulate specific tasks and indicate how the necessary answers will be generated, e.g. through interviews, market research or desk research. The answers are formulated in final deliverables or business model elaborations, which eventually provide insight into possible ways of formulating the business model as well as the critical design issues. In addition, deliverables lead to structural specifications, which cover earlier specified requirements and assumptions. Figure 15.1 contains an overview of the various steps in the design process that was used in the case of the MusicBox service.

Although the figure presented above suggests that designing a business model consists of consecutive phases, it is actually a continuous iterative process. To obtain initial input for the MusicBox business model, we began by conducting interviews and discussions with partners, which resulted in a first sketch of the business model, comparable to the result of the quick-scan described in Chap. 5. Considering this initial sketch, the use of the STOF model made it possible to identify the business model's 'blind spots', i.e. the CDIs help identify the data needed to arrive at sensible business model design choices. Market research, and the way to define a correct marketing strategy, could be useful. Gathering information

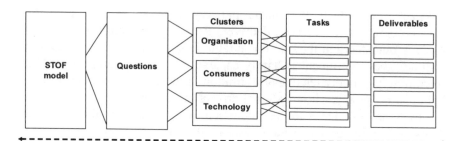

Fig. 15.1. Schematic representation of the design process

regarding market structure, user needs, competitive offerings, technology costs, etcetera, is important for assumptions and requirements regarding targeting, pricing and other design choices, which are part of the service and technology domain. The market research we performed consisted of four phases, i.e. focus group discussions, user-testing sessions, product labs, and in-store end-user interviews.

Focus Group Discussions. The first phase, i.e. focus group discussions, helped us gather information regarding the behavior of end-users. Focus group discussions consisted of discussions among 8–10 individuals with roughly similar backgrounds as well as involvement in music. In addition, a professional discussion leader chaired the discussions, which took place at a secluded location. In general, the focus group discussions provided us with an exploratory understanding of the customer, his or her entertainment space and music habits. Moreover, the focus group discussions elicited some preliminary reactions with regard to the MusicBox concept. As a result, the focus group discussions lead to a first orientation on segmentation and targeting. The STOF model provided guidelines in order to define the 'what' issues that had to be addressed.

User Testing Sessions. The second phase consisted of a user testing session, the aim of which was to obtain consumer reactions to an interactive mock up, in other words an incomplete form of a prototype which is used to make discussions about the functional and the users' requirements less abstract (Verschuren & Hartog, 2005). Accordingly, the user testing sessions helped define the business model specifications. The feedback led to a number of practical (technical) adjustments as well as providing insight into the behavior of potential end-users. Because the STOF method, in addition to guidelines, also contains possible design choices, the feedback from the user testing sessions provides some initial verification and validation with regard to STOF-related issues, such as *Targeting, Creating Value Elements*, technical design choices and *Pricing*.

Product Lab. In the third phase, we used a product lab to obtain spontaneous and realistic consumer reactions to the MusicBox in a simulated environment. Compared to the User Testing Sessions, the product lab involved a more functional prototype and implied a more structured verification and validation of MusicBox. Because the product lab sessions included 8–10 respondents with similar backgrounds, allocation based on the backgrounds of groups is possible. However, the product labs made it possible to validate CDIs as well as verify pricing strategies. Moreover, they helped make it clear to what extent the technological design manages to take on board user requirements and specifications. In addition, the final round of the product lab consisted of a survey. The survey research contains self-completion questionnaires aimed at capturing spontaneous individual reactions, which are often lost in a group scenario. With regard to the technical part of the prototype, the STOF model focuses on the design of the technical architecture at a high level. Consequently, in the product lab phase the STOF model is more suitable to evaluate CDIs from the service domain and various pricing strategies. Having said that, the evaluation of the CDIs from the service domain typically influences the CDIs from the technology domain, as the latter set of CDIs enable the first.

End-User Interviews. The fourth phase consisted of (in-store) end-user interviews, the aim of which was more or less the same to that of the product labs. As such, the end-user interviews also resulted in the validation and verification of value elements and pricing. Although the end-user interviews were part of the final market research phase, real 'beta testing' will take place during the pilot.

15.4 Service Domain

Within the service domain, the *Value Proposition* is the central issue. The value proposition of MusicBox defines its intended value. Any innovative service such as MusicBox can only be successful when it addresses market needs and manages to realize sufficiently high levels of adoption. Therefore, the service domain of the business model specifically focuses on end-user value aspects (CDIs), such as *Targeting, Creating Value Elements, Branding* and *Customer Retention*. Needless to say, the specific characteristics of the MusicBox have to be taken into account. One of the issues has to do with the location of MusicBox, which will significantly contribute to the value of the service. In addition, the issue of piracy is a typical characteristic of the music industry in industrializing countries.

Table 15.1. CDIs with regard to the MusicBox service in the Service domain

Critical design issue	Balancing requirements	Present in MusicBox	Notes
Targeting	Generic vs. niche service	++	Based on disposable income, high-tech services mainly consist of niche services
Location selection	Widespread vs. dense deployment	++	Because MusicBox requires a fixed location, the selection of optimal locations is a CDI related to the generic CDI *Accessibility for Customers*
Creating value elements	Technological possibilities vs. user needs and wishes	++	Based on disposable income, most services in this country meet the needs and wishes of the end-user rather than providing advanced technologies, which is exactly the case with MusicBox
	Ease of use vs. added value		A balancing requirement is implied by ease of use vs. added value. An increased level of experience with MusicBox could generate an interest in additional features, e.g. novice vs. experienced user
Branding	Distributor (lifestyle or device brand) vs. content brand	+	Although the CDI of branding is significantly different from that of mobile services, branding is important for creating value around the service. However, because branding ideally implies a strong lifestyle brand, it has to be a 'core' partner
Customer retention	Customer lock-in vs. customer annoyance	++	Because the NFC smartcard is a prepaid payment solution, the card provides an initial lock-in. However, balancing was visible in the discussion surrounding the marketing strategy, and a choice was made in favor of a loyalty program to avoid customer annoyance

End-user education	End-user instruction vs. independent use	++	Although the issue of end-user education is not explicitly part of the STOF model, as far as the success of MusicBox is concerned it is a crucial issue that is related to the generic CDI *Accessibility for Customers*. Therefore, in the case of MusicBox, end-user education is enclosed as a CDI. It could be a characteristic of markets in newly industrialized countries

Generally speaking, the CDIs provided by the STOF model are applicable in the case of MusicBox. However, to establish a better fit, the CDIs needed to be somewhat adjusted with regard to the balancing of requirements of *Branding* and *Creating Value Elements*, when compared with the balancing of requirements as suggested in Chap. 3. Table 15.1 illustrates the various CDIs along with the preferred requirement, as well as the degree to which the various issues are present in the MusicBox case (++ = strongly present to –– = not present at all).

Besides the CDIs offered by the STOF model, two additional CDIs are applicable: location selection and end-user education. Both are examples of the generic CDI *Accessibility for Customers*. The location of the service fairly literally determines its availability to customers. Users need to be educated on the use of the service and be taught the required skills they need to access the service. Furthermore, branding MusicBox implied some different balancing requirements compared to what is usual for mobile content services. In addition, the arguments in favor of the choices that were made with respect to the balancing requirements of *Targeting*, *Creating Value Elements* and end-user education could indicate factors that are characteristic of markets in newly industrialized countries. Even though the STOF model addresses the issue of targeting, a subdivision in segmentation and targeting could lead to a more dynamic approach in the sense that the target group may change during the service's lifecycle.

Critical issues with regard to the commercial success of MusicBox include *Targeting*, and *Market Segmentation*. However, the issue of *Targeting* involves more than merely defining the MusicBox target group; it also addresses the context in which we ought to see MusicBox. In addition, *Targeting* implied identifying the individual needs, behavior, wishes and interests, in other words, the user requirements, within end-user groups, and defining the appropriate personal or social context. Another issue that is of decisive importance in relation to *Targeting* is the *Accessibility for Customers* with respect to MusicBox. Decisions regarding these issues are important in determining the expected size of the target group and predicting the percentage of end-users.

15.5 Technology Domain

The technology domain addresses the requirements generated by other domains. Consequently, this domain creates a technical architecture based on the requirements and technical CDIs, such as *Security, Quality of Service, System Integration, Accessibility for Customers* and *User Profile Management*. In general, the CDIs provided by the STOF model are roughly applicable in the case of MusicBox. Even though the CDIs of standardization and scalability are not marked as 'critical' in the STOF model, they are essential CDIs in the case of MusicBox, which is why they are added to Table 15.2, which provides an overview of the various CDIs, together with the preferred requirement and the degree to which these issues are present in the MusicBox case (++ = strongly present to — = not present at all).

Table 15.2. CDIs for the MusicBox service in the Technology domain

Critical design issue	Balancing requirements	Present in MusicBox	Notes
Security	Ease of use vs. preventing abuse and piracy	0	An important security issue with regard to digital content is piracy. However, because adoption levels of DRM are low in these countries, ease of use is applicable
Quality of service	Quality vs. costs	++	The QoS balancing requirement is constantly applicable. Accordingly, Company A prefers quality above costs and asserts that end-users are willing to pay for quality
System integration	Flexibility vs. costs	++	System integration is a continuous factor during the design of the technical architecture. Since technology advances rapidly, flexibility is very important for a service like MusicBox. Also, the service needs to be market

			proof for the next 5 years, providing all kinds of digital content
Accessibility or compatibility	Open vs. closed system	+	In the case of MusicBox accessibility is an issue involving compatibility (standards). Because Nokia is a partner, the Sony Ericsson phones are not compatible. However, the service is available to anyone with a compatible device
Management of user profiles	User involvement vs. automatic generation	++	The balancing requirement leads to a discussion as to the kind of basic information that is required for the user ID. In addition to nickname, age, and gender, MusicBox will automatically generate a profile. The profile allows for personal recommendations and is also used to register end-user behavior
Standardization	Innovation speed vs. similarities between standards	++	Even though standardization is not considered a CDI for mobile services, in the case of MusicBox it is essential for the development of the service. The issue describes how accepted standards should be used. In addition, Company A prefers innovation speed, because time to market is very important for MusicBox
Scalability	Costs and uncertainty of careful planning vs. learning as going along	++	In light of the fact that Company A plans to roll out MusicBox in other countries as well, provided it is successful, scalability is a very relevant CDI. The issue describes how and to what extent the technology is scalable. In view of the scalability of the technical architecture, a distinction between scale-out and scale-up should be made

Although security issues are applicable in the case of MusicBox, they are not critical to the design, because there is no anti-piracy regulation. In addition, because the number of players in the market that enable protection of digital music is still low, security only plays a small role, which means that it is a CDI, but not a prominent one. The opposite is true for scalability, which is a CDI that is 'strongly present' in the MusicBox case. Scalability can be defined as the ability to scale up (vertically) or scale out (horizontally) in case the service needs to expand (Devlin, Grey, Laing, & Spix, 1999). As far as MusicBox is concerned, technological, economic, and geographical scalability are important. The technology domain only addresses technological scalability, while in the STOF model the other two kinds of scalability are addressed in the organization domain. Technological scalability refers to the ability on the part of MusicBox to increase the total number of digital content transfers when hardware is added. Technical scalability is important when the number of end-users increases or MusicBox starts offering different forms of digital content in addition to high quality music.

When the number of end-users increases, there is a need to scale out (horizontally), which implies adding more MusicBox units to the system, for example by linking multiple MusicBox units to one broadband connection, payment mechanism or service engineer, or by connecting more than one unit to the same payment gateway, central server or music database. If MusicBox starts offering other services or types of content, it needs to scale up (vertically), which implies adding resources to a single vending unit in the system, i.e. adding memory, interdependencies, and applications. Scaling up MusicBox is necessary when Company A decides to install multiple screens supported by a single unit, i.e. one computer, connection and note collector. Other examples involve the connection and support of more than one device at a time or providing different services, i.e. different digital content, through MusicBox.

15.6 Organization Domain

Next, the organization domain of the business model focused specifically on organizational network aspects, which includes CDIs like *Partner Selection, Network Openness, Network Governance* and *Network Complexity*. These aspects revolve around a value network of actors, which indicates that there are specific partner roles and resources. Although Company A has some experience with regard to *Partner Selection*, not all partnerships have been successful. As a result, selecting partners was a meticulous process.

Moreover, because the synergy among the partners to a large extent determines the potential success and quality of the service, *Partner Selection*, partnership scalability and entry rules became important issues.

Even though the STOF model provides criteria on which the partner selection can be based, size or scalability are additional criteria. Also, the issue of trust, as well as earlier experiences from other partnerships, influenced partner selection within the MusicBox network. Available resources in the STOF model indicate the 'kind' of resource and not the quantity, which is an important issue with regard to scalability. Table 15.3 illustrates the different CDIs in the organization domain, along with the preferred requirements, as well as the degree to which these issues were present in the MusicBox case (++ = strongly present to — = not present at all).

Table 15.3. CDIs for the MusicBox service in the Organization domain

Critical design issue	Balancing requirements	Present in MusicBox	Notes
Partner selection	Limited number of partners vs. quality of service and strategic interests	++	Even though this CDI is highly important, the balancing requirements are not necessarily a balance, because a limited number of partners can also result in a higher quality of service and be a result of strategic interest. In the case of MusicBox, both kinds of balancing are applicable
Network openness	Openness and customer reach vs. control and exclusiveness	++	Network openness is an essential design issue, as the arrangements regarding network openness define the boundaries of the service
Network governance	Entry, compliance and exit conditions: individual vs. network interest	+	Even though this issue is applicable, entry compliance and exit conditions are not clearly identified in the initial phase of development
Network complexity	Need to reduce complexity vs. need of access to critical resources and capability's	++	For each part of the design or value network one deploys either a single organization or multiple organizations. Therefore, it is a continuous issue

Even though the organizational CDIs in relation to MusicBox are distinctly present, the CDI of *Network Governance* is less clear in the initial phase. As a result of the relatively unclear initial agreements, the conditions and balance to a large extent remained vague. Nevertheless, the CDI became more clear and critical as the design process progressed.

The core of the MusicBox value network, basically a 'walled garden' network, consists of the end-user and the most important partners, i.e. content providers and aggregators, middleware and service provider, and distributor. Some factors did lead to the decision to work together with specific partners. These factors included, among other things, the available resources and competences as well as the existence of direct or indirect relationships with a company. The latter aspects together created trust among the partners in the value network, which is one of the main requirements for a network to function. Besides, issues such as strategic fit and shared objectives can be decisive in partner selection. Additionally, the current market position and market role of a company or brand can provide a reason for including it in the partner network. Key drivers with regard to partner selection in the case of MusicBox were size and scalability. Scalability to a large extent depends on the partners that are assimilated into the partner network.

In addition, there is a strong link between the arrangements within the partner network and the financial arrangements. Moreover, that link not only influences the decisions in the other domain, it appears to be practically integrated within certain decisions (e.g. *Network Governance*, and *Division of Investments* and *Valuation of Contributions and Benefits* between network actors). Referring to Chaps. 3 and 5, we see that the organizational and financial arrangements together are instrumental in balancing and arranging actors' interests.

15.7 Finance Domain

As far as the development and exploitation of the service was concerned, design decisions in the finance domain were closely related to decisions regarding the organizational arrangements. Accordingly, the finance domain addresses financial CDIs, such as *Pricing* and *Division of Investments*, as well as the *Valuation of Contributions and Benefits* between network actors. The price of the content will be crucially important in terms of the adoption and success of the MusicBox. The core partners in the value network share revenues on a royalty base. Pricing is part of the service domain as well as the finance domain. Table 15.4

Table 15.4. CDIs for the MusicBox service in the finance domain

Critical design issue	Balancing requirements	Present in MusicBox	Notes
Pricing	Maximizing profits vs. creating market share	++	Pricing is a highly important critical design issue for the service. Ultimately, the price of content strongly influences whether or not end-users are willing to start adopting the service
Division of investments	Operational financial interest (ROI) vs. intangible benefits (options)	++	All the partners invest based on different reasons, and they participate with dissimilar interests
Division and valuation of costs and revenues between network actors	Costs-benefits valuation on level of network vs. cost-benefits for individual partners	++	Two different mechanisms for the valuation and the division of costs and revenues are used. On the core network level (tier-1) division is characterized by royalty sharing. For individual suppliers (tier-2 and -3) fixed revenues are agreed upon

contains the CDIs for the finance domain along with the preferred requirements, as well as the degree to which these issues were present in the MusicBox case (++ = strongly present to —— = not present at all). The two generic CDIs for *Division of Costs and Revenues*, and *Valuation of Contributions and Benefits*, respectively, are taken together.

The CDIs for the finance domain are all 'strongly present' in the case of MusicBox. However, the *Division and Valuation of Costs and Revenues* between the network actors indicates different design choices on the different network levels. At the level of the core partners (tier-1), revenue sharing is the preferred choice. At the level of individual suppliers (tier-2 and -3), fixed revenues are agreed.

For a business model to be viable, the service needs to generate enough revenues. For a stable business model, it is also important to ensure that costs, risks, and investments are divided in such a way as to be acceptable to all parties involved. Keeping the critical success factors in mind that were discussed in Chap. 3, this means that the service should be profitable to the network. Accordingly, the division of content-related revenues among the tier-1 partners results in a royalty sharing agreement. As a result, the risks involved were shared by the tier-1 partners, and as a result the trust among the partners in the network increased. Apart from taking into account the various levels of risk involved, the tier-1 partners pay suppliers based on fixed prices.

The price of content will be crucially important to the adoption of the MusicBox, i.e. whether the service generates sufficient customer value. However, determining the optimal price based on market research remains difficult, because there is no proper benchmark for the price of (legal) digital music. Nevertheless, it will be easy to adjust the pricing system of MusicBox during the initial market roll-out in order to determine the optimal price for the service content.

15.8 Conclusion

The STOF model proved to be useful for designing the MusicBox service and the underlying business model. The STOF model and method make it possible to structure the complex set of issues that have to be dealt with when designing a service. It proved to be a practical framework for creating insight into the various issues and interdependencies. In this chapter, we presented a typical way to apply the STOF model and method. In the case of MusicBox, a digital service to be introduced in a newly industrialized country, the STOF model and method provide clear guidelines with regard to the first phase of business model design, i.e. the phase that deals with the design on paper. As a result, the guidelines cover most of the important generic issues, components and choices. In addition, based on the guidelines it is possible to create an overview of the various components of a service business model. Furthermore, the STOF model helps gain insight into the interdependencies between the various components, making it possible to understand the influence that a decision involving one component has on other components or even domains. Therefore, it becomes easier to see how the business model should be adapted to accommodate changes in, for example, the target group, service location or geographic region. In addition, the STOF model helps us understand the scale dimension, and provides support when switching to a different revenue model or new technology. Because the STOF model addresses broad and generic issues, the approach is applicable to various services.

In this case there was a need to interact with various layers within the organization. The STOF method does not provide explicit guidelines for such situations. Typically, the STOF method is applied to situations involving people who operate at the same organizational level. Applying the method at lower levels is too complex, in which case developing detailed process models for a specific area is an obvious approach.

Targeting is one of the critical design issues with regard to the service domain. As far as MusicBox was concerned, it was important to have a subdivision in segmentation and targeting, which leads to a more dynamic approach. Because defining the target group involves making a static observation at a particular moment, market segmentation makes it possible to plan the best way to roll out a service. In addition, including more guidelines on customer (end-user) acquisition and typical interdependencies within the various STOF domains could improve the prescriptive value of the model. Other issues that turned out to be relevant to MusicBox and are not included in the STOF model have to do with selection of locations and end-user education.

Chapter 16 From Prototype to Exploitation: Mobile Services for Patients with Chronic Lower Back Pain

E. Fielt, R. Huis In't Veld, and M. Vollenbroek-Hutten

Many research and development projects that are carried out by firms and research institutes are technology-oriented. There is a large gap between research results, for instance in the form of prototypes, and the actual service offerings to customers. This becomes problematic when an organization wants to bring the results from such a project to the market, which will be particularly troublesome when the research results do not readily fit traditional offerings, roles and capabilities in the industry, nor the financial arrangements.

In this chapter, we discuss the design of a business model for a mobile health service, starting with a research prototype that was developed for patients with chronic lower back pain, using the STOF model and method. In a number of design sessions, an initial business model was developed that identifies critical design issues that play a role in moving from prototype toward market deployment. The business model serves as a starting-point to identify and commit relevant stakeholders, and to draw up a business plan and case.

This chapter is structured as follows. We begin by discussing the need for mobile health business models. Next, the research and development project on mobile health and the prototype for chronic lower back pain patients are introduced, after which the approach used to develop the business model is described, followed by a discussion of the developed mobile health business model for each of the STOF domains. We conclude with a discussion regarding the lessons that were learned with respect to the development of a business model on the basis of a prototype.

16.1 The need for Mobile Health Business Models

From literature it is known that the majority of e-health services do not make it to full deployment (Tanriverdi & Iacono, 1999). According to Berg (1999) more than 75% of the e-health services should be considered operating failures. One of the main reasons for this failure is the fact that it is not known how to make these services financially viable as there is insufficient knowledge about their business models. In this section we discuss the need for mobile health business models.

The capacity of our current healthcare system is insufficient to continue treatment of patients in the conventional way involving face-to-face (called in vivo) treatments. This capacity problem is caused by the growing number of elderly (i.e. aging) people, combined with a reduction in number of healthcare professionals (*VWS*, 2004). As a consequence of the aging of our population, the number of chronic diseases is growing as well, which further increases the need for healthcare, in a setting where fewer resources are available. This means that new, more effective and more efficient ways of treatment need to be developed and deployed.

The advances in ICT capabilities and the availability of mobile devices and high bandwidth public wireless networks create possibilities for new mobile healthcare services, including remote monitoring and treatment for chronic diseases. These services are expected to be more efficient, because individual therapists can treat several subjects simultaneously anytime and anywhere, making it possible to replace costly intramural care by less costly extramural care. These services are also assumed to be more effective because patients train in their own home or work environment, and they are not limited to the availability of the therapist, but can train much more intense. The level of compliance and practice is believed to be a key factor in obtaining results in (physical) therapy.

As a result, research and development involving new e-health applications is rapidly growing. However, as mentioned earlier, e-health initiatives often fail after the funding phase. The difficulties with large scale deployment of e-health technologies may, among other reasons, be explained by the variety of factors that influence technological innovation in the health sector. The factors that affect the deployment of new e-health applications can be divided into five types: (1) behavioral (acceptance, diffusion and dissemination), (2) economical (policy and legislation, standardization, security), (3) financial (provider, structure), (4) technical (support, training, usability, quality of service), and (5) organizational (overlap with existing work practices) (Broens et al., 2007; Tanriverdi & Iacono, 1998). To understand actual deployment specific, problem-related knowledge on each of the five areas is needed.

However, even when domain-specific expertise is available, it needs to be bundled and integrated. Because there is a lack of 'shared vision' among all the stakeholders involved (for example with regard to the division of roles, costs, revenues, et cetera), implementation remains difficult. The situation is particularly complex in the health domain, where the relevant tasks and responsibilities are divided among different parties, including doctors, hospitals and insurance firms. Moreover, e-health also involves techno-logical stakeholders in the development and delivery of the health service. Most new e-health initiatives focus on technical and clinical capabilities, without taking the service production and consumption, relevant stake-holders, financial arrangements and organizational issues into account. Developing a business model helps create focus and addresses these issues in a systematic way, facilitates communication amongst different stake-holders, and results in a shared helicopter view of the initiative.

16.2 The Prototype for Chronic Lower Back Pain

In this section, we introduce the prototype application for chronic lower back pain that has been developed within the Awareness project. The goal of this project is to research and design a service and network infrastructure for context-aware and pro-active mobile applications, and to validate these through prototyping with mobile health applications. Context-related information refers, for example, to the location and availability of people. Within the Awareness project, Roessingh Research and Development (RRD) initiated and leaded service development for patients with lower back pain for which a prototype was developed (Van Weering, Jones, & Thiele, 2007). RRD is a research institute in the area of rehabilitation medicine, in particular rehabilitation technology and pain rehabilitation.

The service concept of the prototype focuses on balancing daily activity patterns of patients in order to reduce chronic lower back pain. The mobile health service is aimed at the long-term monitoring of physical activity in the patient's own environment, and at providing feedback to the patient about deviating activity levels in relation to healthy controls. There are three types of feedback: standard (e.g. have a rest), personal (e.g. have a cup of tea – for those who like tea), and personal with context (e.g. have a cup of tea in the garden – when the weather is nice and the person is at home). Figure 16.1 provides a schematic overview of the prototype. The prototype will be described in greater detail when we discuss the technology domain of the business model.

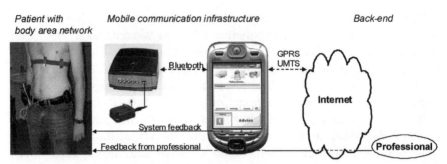

Fig 16.1. Schematic presentation of the service concept for chronic lower back pain patients (Van Weering et al., 2007)

16.3 Business Model Development

Based on the prototype, a business model was developed that was aimed at mobile health services for chronic back pain patients (Huis In't Veld, Fielt, Faber, & Vollenbroek-Hutten, 2007). The business model should increase our understanding and enhance the viability of the mobile health service. A business model can help moving a service from the research phase toward the implementation phase, and serve as a basis for a business plan and business case as explained in Chap. 2. This is particularly relevant for RRD because of their explicit objective to implement their knowledge in rehabilitation practice and the development of rehabilitation products.

To develop the business model, the STOF model and method were used. This chapter presents the development of the initial business model from an internal project-oriented view. This initial model will be further developed and improved with external stakeholders. To develop the initial business model a sequence of six design workshops of about 2 h each was organized between June and September 2007: two workshops for the service domain, one for each of the other domains, and a wrap-up and integration workshop to conclude the development.

The workshop participants were RRD members (both researchers and staff), since RRD is the primary stakeholder for the business model development, extended with researchers of the Awareness project. The participants came from different backgrounds, including physiotherapy, technology, finance and management. For financial knowledge a person from a Dutch insurance company was invited. Each workshop started with an update of the business model as it has been developed so far. Whenever it was deemed necessary, open issues that arose during a workshop were investigated in greater depth afterwards.

Below, we provide a description of the various workshops by exploring the service, technology, organization and finance domains.

16.3.1 The Service Domain

The prototype for patients with lower back pain is a tele-health service. In general, the advantages of tele-health services are that they decrease the need for the patient and the professional to be at the same place at the same time, a less direct interaction between patient and professional, depending on the extent to which the services are automated.

The various possible (generic) tele-health services (Fig. 16.2) and their value propositions were discussed in the workshops. The focus of the prototype was addressed in relation to these tele-health services. Basically, the prototype is a tele-monitoring service that makes it possible to gather more health information (more detailed and over a longer period) and information of a better quality (during normal activities in the user's own environment). This tele-monitoring service enables tele-treatment, i.e. providing the patient with feedback either automatically via a device or personally via a professional. In general, tele-monitoring and tele-treatment can increase the quality and efficiency of health services. The specific benefits depend on the extent to which tele-health services are used to replace and/or complement traditional health services. Tele-monitoring service also enables tele-alarm by using the gathered information to detect unexpected patterns that trigger an automatic alarm for patient and/or assistance. Tele-consultation and tele-community are additional possibilities, with or without tele-monitoring. In addition, there are (tele-)advice and support services for patient and professional.

For the chronic lower back pain application, a combination of tele-monitoring and tele-treatment are the core services supplemented with tele-consultation services and advice and support services. Tele-community is a possible additional service, while tele-alarm is not relevant for chronic lower back pain.

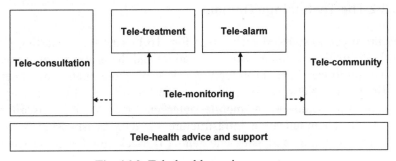

Fig. 16.2. Tele-health services spectrum

An obvious target group for the chronic lower back pain application consists of patients with chronic lower back pain and the professionals who treat them, such as physiotherapists. However, it makes sense to identify other potential markets in order to increase the overall market size. This creates opportunities to (partially) leverage the investment in the application and infrastructure and to mitigate the risks when the acceptance levels of the service or the cooperation of the stakeholders in this market turn out to be disappointing. An additional target group consists of employees with lower back problems and the occupational health providers responsible for their treatment. Since the prototype focuses on influencing activity patterns, a third target market was defined that consists of consumers with obesity, hence focusing on the wellness market as well. In this market, the activity pattern can also be an important part of the treatment.

The three target groups were selected because of their similarities concerning monitoring and treatment. However, there are also differences as far as this particular service is concerned. For example, patients and employees have a higher need for increasing the range of sensing (e.g., physical condition, hearth rate) and require more frequent and detailed feedback. Together, these three markets cover the individual and business segments of the healthcare market as well as the consumer market. These markets differ with respect to the organizational settings and financial arrangements, as we will see when we discuss the other domains.

Specific attention in the service domain description was paid to an idealized, future context-of-use for mobile health services. This involved covering a range of wellness and healthcare issues via 'plug and play' sensors and services with one central device and tailored to the person and context-of-use. Preferably, the device is someone's own mobile phone. While the chronic lower back pain application is still far removed from this ideal future, it is very important to take the ease-of-use for the patient into account.

16.3.2 The Technology Domain

The prototype introduced earlier (see Fig. 16.1) serves as a starting-point for the technology domain. Both the patient and the professional are users of the prototype. The prototype consists of three main components: (1) Body Area Network for patients, consisting of an activity sensor, an actuator for feedback, a mobile interface and medical algorithms, (2) Mobile communication infrastructure for data transmission between body area network and back-end, and (3) Back-end for professionals, including user management, storage/back-up facilities, medical algorithms, medical display, tele-video consultation and context management.

The existing prototype can be conceived as a high-end solution. Offering basic tele-monitoring and treatment may also be possible with a simpler communication infrastructure (e.g. not mobile or not continuously) and back-end, and with a lower level of (or no) personalization and context-awareness.

In general, the technological solution may vary with respect to the extent to which the computer can support activities, and based on the question whether the device or the (mobile) network should be the core of the solution. Figure 16.3 presents four possible solutions, focusing on the role of the patient's device. In the case of the 'smart device', the mobile health service is a stand-alone service on a device (e.g. automated feedback). When remote processing and storage are required, the 'connected device' is available. When human involvement is necessary (e.g. professional feedback), the patient can take the device to the professional (shared device) or the professional can access the device remotely, anytime and anywhere (real-time device). The increasing capabilities of the device (e.g. processing power, battery duration, and storage capacity) enhance the possibilities of the smart device. However, there are also developments that make the other solutions more attractive, for instance a demand for small devices or cheaper mobile data transmission.

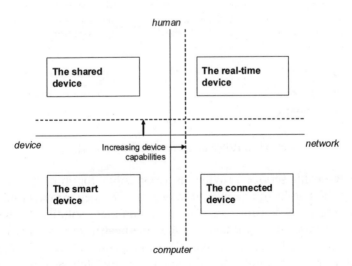

Fig. 16.3. Solution space for tele-health devices

There are potential technological risks that have to be taken into account. The prototype's Body Area Network requires a certain number of devices per person and some devices are quite expensive. Reducing the number of devices would make the system cheaper and easier to use. Power consumption is a problem when continuous real-time monitoring over a long-period is required and processing and data transmission requirements are high. When activity sensors become less exclusive, it may be easier to introduce alternative solutions. In particular the obesity version of the service may well function with a simpler and, therefore, cheaper and widely-available sensor. Finally, there is a lack of standards regarding (context-aware) tele-health services at a national and international level, which may have a negative impact on the level of acceptance in the regular healthcare sector.

16.3.3 The Organization Domain

The tele-health network aims at offering tele-health services, where the following roles can be identified: tele-health service provider, demand-side roles, and supply-side roles (Fig. 16.4). The tele-health service provider is the focal role responsible for implementing and delivering the tele-health service, bridging the health sector on the demand-side and the technology sector on the supply-side.

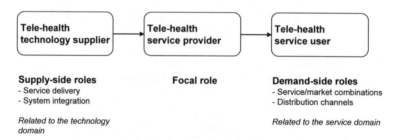

Fig. 16.4. Roles in the tele-health network

The tele-health service provider is responsible for implementing and delivering the tele-health service to the users. This new provider takes the initiative and is responsible for offering a complete solution to the user. Activities (possibly) performed by the tele-health service-provider are managing the tele-health initiative, developing the relations in the tele-heath network, arranging the financing of the tele-health initiative, handling legal issues for the tele-health service, marketing and sales of the tele-health service, responsible for delivering the tele-health service, user training and support (both end-user and health service provider), and stimulating research and development for tele-health services.

The tele-health service-provider can perform fewer activities when the activities are or become available as services in the market, for example via partnering with an existing tele-health service-provider. The tele-health service-provider can also perform more activities related to service delivery and system integration, discussed in the supply-side roles. The exact activities of the tele-health service provider will depend on the kind of actor playing that role and the strategic choices of that actor, as later discussed in the start-up scenarios.

The demand-side roles vary for each potential target market. The end-users are the patients, employees and/or individuals suffering from obesity (consumers). As far as chronic pain is concerned, a health provider (a health professional or health organization) is involved, while in the case of obesity agents (e.g. fitness centers) and direct service sales and delivery offer alternatives. With regard to work-related health, an occupational health provider is involved. The (occupational) health provider integrates the tele-health service into the overall health service involving chronic or work-related lower back pain. Moreover, a tele-health portal, acting as a single source for different tele-health services, can be involved, offering, for example, neck shoulder pain tele-health next to lower back pain. The distribution choices will affect the activities of the tele-health service provider, such as marketing and sales and training and support.

With regard to the supply-side roles, we separate the service delivery from the system integration. The tele-health service provider has to arrange the actual service delivery with a tele-health platform operator and with data, communication, and context service providers. As far as the platform and device are concerned, system integrators are needed, who use software and hardware suppliers, and (medical) R&D organizations. With regard to the roles of hardware and software supplier, it may be useful to differentiate between those that provide hardware and software that is specific to the tele-health service and those that provide hardware and software that is more generic in nature, for instance mobile service platforms or personal devices.

One of the major issues is the question as to who will be the central actor playing the new role of tele-health service provider (and portal). We present four possible start-up scenarios:

1. *Demand-Side Scenario.* Actors involved in the demand-side of the tele-health network, such as health providers or health insurers, can take on the role of tele-health service provider as part of their own organization or invest in a separate organizational entity.

2. *Supply-Side Scenario*. Actors involved in the supply-side of the tele-health network, such as health solution providers (e.g. Philips) or communication providers (e.g. KPN), can take on the role of tele-health service provider as part of their own organization or invest in a separate organizational entity.
3. *Outside Investor Scenario*. Actors who play no specific role in the tele-health network, such as venture capitalist, can invest in the start-up of a new organizational entity that is directed towards the role of tele-health service provider.
4. *Mixed Scenario*. This scenario is a combination of the other there scenarios. Different actors cooperate and together start-up a new organizational entity that is directed towards the role of tele-health service provider.

For each of the scenarios the central role is allocated to an actor willing to invest in a new enterprise. These investment issues are discussed in the finance domain).

16.3.4 The Finance Domain

With regard to the finance domain, we discuss three possible revenue sources. The first identified revenue source involves health insurance, most suited to chronic pain service, while there are also some possibilities for obesity and work related pain. The second source is employer insurance, which is targeted at employers, and this is most suited to the work-related lower back service. With insurance as revenue source, the costs for using the service are paid for by an insurance company. The third is commercial sales (where users purchase a device or subscribe to the service), which is most suited to the obesity service. Table 16.1 provides an overview of the matching between the services and revenue models.

Table 16.1. Revenues sources for the tele-health services

	Chronic pain	Work-related	Obesity
Health insurance	++	+	+
Employer insurance		++	
Commercial sales			++

'++', a very good fit; '+', a good fit between revenue sources and target group

With regard to the insurance models, we start with a number of assumptions about the future development of the healthcare sector, based on existing trends within the Dutch healthcare system discussed during the workshop. There is an incentive to adopt healthcare innovations due to the need for productivity improvement because of the expected lack of capacity. Moreover, it is expected that there will be a shift from more budget-driven financing toward more market-driven financing. It is to be expected that even the threat of introducing a market model in healthcare may result in a change in the healthcare system. Market-driven financing means insurers will pay for the outcome of the treatment, not for the activities involved. Health professionals/organizations (for health insurance) and occupational health providers (for employer insurance) will compete for outcome contracts (fixed price) with health insurers.

The health professionals and occupational health providers have incentives to improve their health services because of the outcome contracts and because of mutual competition. As a result, they will start using tele-health services when it offers benefits to their health services. To use the tele-health services, the health professionals and occupational health providers need a contract with the tele-health service provider. This also means that the tele-health service provider is paid by the health professional rather than the health insurant or insurer. The benefits that tele-health services offer to the health services of health professionals and occupational health providers offer possibilities for a higher quality of service (more effective, more easily accessible, higher availability) against lower costs. The expected benefits will depend on the extent to which the tele-health service complements or replaces parts of the traditional health service. Moreover, realizing the benefits depends on whether the necessary social and organizational changes take place. For example, will patients accept tele-health services and can health professionals integrate them into their work processes? The benefits of tele-health services to health professionals and occupational health providers can result in higher margins and/or more competitive prices.

In the case of obesity, there is the possibility of selling commercially to consumers on an individual or periodical (subscription) basis. When the tele-health service provider deploys agents, these agents will expect to receive a commission. The consumer price can be lowered by sponsoring via health (or non-health) insurers, because it can prevent insurance claims in the future and/or because being associated with a tele-health service can provide insurers with an innovative image. Future insurance claims can be related to psycho-social and medical (e.g. heart diseases and diabetes) complaints related to obesity.

The tele-health service provider's revenues and costs are based on providing the service to the three target markets. The revenues come from the contracts with health professionals/organizations and/or occupational health providers and/or commercial sales to consumers, most likely in the form of a subscription fee that can be a combination of a fixed amount, an amount per user/device and/or a usage-related charge (e.g. number of hours, amount of data). The tele-health provider may differentiate in price for a basic service and a top-up for more luxurious options (personalization, context-awareness). Promotional and quantity discounts can decrease the revenues.

The costs of the tele-health service provider have to do with delivering the tele-health service and running a business in general (e.g. marketing, housing, administration, et cetera). The kind and number of devices and sensors will affect the costs of delivering the tele-health service. Re-use of devices and sensors can reduce the costs, especially with regard to the patient and employee market. The tele-health service provider may also need to invest in hardware and software for the health professionals/ organizations and occupational health providers. Another important cost factor is the extent to which the services for the three target markets are able to make use of the same processes and systems. While this is the intention, in the end it may not be feasible when the treatment and the devices/sensors are too different. An important cost-related trade-off is that between investing up front in ease-of-use and reliability, and having higher training and support cost (in case of service break-downs).

16.4 Conclusion

In this chapter, we illustrated the use of the STOF model and method to develop a mobile health business model starting from a prototype aimed at patients with lower back pain. The STOF model and method helped preparing and conducting the workshops, and organizing and presenting the results (Huis In't Veld et al., 2007). The presentation of the results in a feedback session to the project members gave the impression that the model and method address relevant issues in a systematic way. We experienced that organizing separate workshops for each of the domains provides more time for discussion, although it also makes it harder for the participants to get an overall picture (due to the time that elapses between workshops) and feel committed to the overall result (due to changing participants per workshop). Therefore, it may be advisable to start with an overall workshop, before moving on to separate workshops for the individual domains.

The development of the business model resulted in a change from a focus on the technology toward an overall picture that including the service, organization and finance domains. Moreover, it resulted in a rethinking of the technology choices that were made while developing the prototype. The service domain opened up the discussion regarding a use of application to serve other target groups and to take additional services into account. It also became clear that there is a need for a more concrete and elaborated description of the service, and that this requires more interaction with and knowledge about the target markets. The organization domain draws attention to the new role of the tele-health service provider and the need for a new or existing actor to take up this role. In addition, the organization domain makes it clear that the initiative requires a complex organizational network that depends on the cooperation of multiple stakeholders with different interests and capabilities. While in general there is agreement that e-health services are needed to keep providing healthcare to a growing population with the same number of fewer health professionals, the finance domain showed that the revenues and costs are not yet very clear. Moreover, the distribution of the revenues and costs among the various actors remains an important issue that needs to be resolved.

The development of the business model also resulted in rethinking the technology choices that were made with regard to the prototype. The existing prototype is a high-end solution that may not be suitable for commercial exploitation, at least not with regard to the basic service or in the near future. The amount and prices of the devices and sensors in the Body Area Network have to be reconsidered, as has the need for mobile communication and context-awareness. Moreover, whether or not technology is a potential barrier for the introduction of mobile health services (e.g. power consumption) also depends on the ambitious claims that have been made about the possibilities for continuous and real-time tele-monitoring.

Chapter 17 From Prototype to Exploitation: Organizational Arrangements for a Personalized Dementia Directory

H. De Vos, T. Haaker, and R.-M. Dröes

One of the typical aspects of applied scientific and technology-driven research is that it focuses on the development of a specific technology that aims at solving a specific problem of a specific target group. In terms of the STOF model, this kind of research aims at specifying the technology domain and parts of the service domain. Attention should also be paid to other aspects, i.e. the organization and finance domains to make the technology available to a broader public at a later stage.

In this chapter we sketch a procedure to broaden the scope of a traditional technology push research project, using parts of the STOF method. To encourage potential business model discussions, we designed several organizational arrangements and discussed them with stakeholders and potential partners. We combined a decision support method making use of the Critical Success Factors to select a business model.

17.1 Introduction

Applied technology research often focuses on solving specific problems that potential users may encounter. In the healthcare domain, this usually involves people who suffer from a certain disease. A common practice in human-centered design is to involve potential users or domain experts in the exploration phase, to identify user requirements and to evaluate the technology in a pilot study. The research is typically carried out in a project with a specified budget for all the activities involved. The hardware and software required are made available for free. Users and other parties are not faced with additional costs and are often willing to cooperate in a pilot. Typically, researchers want users to continue using the product after the research project has finished. However, entering the exploitation phase is not a foregone conclusion. In fact, a large number of prototypes with successful pilots never reach this stage. The step from prototype to market introduction is a big one. Common reasons for projects not to reach the market are that there is no (commercial) party to take the lead in the exploitation phase. Often the aim of these projects is to solve a societal problem, with a variety of providers involved and with users as well as society as a whole benefiting. However, in those cases it is not clear what the benefits are and what the benefits are for the specific provider(s) and the provisioning network as a whole. Among the specific problems in healthcare are how to comply to regulation and to the complex financial structure. As far as the development of exploitation models is concerned, this poses a challenge. A first step towards market introduction is to involve providers and take the providers' interests into account as early as in the design phase. The STOF model and method provide the ingredients needed to realize this, taking into account the interests of the users as well as the providers.

In this chapter we use a Personalized Dementia Directory (PDD) as an example. This PDD provide personalized information on services that fulfill the specific needs of a dementia patient. In the following section, we explain the PDD and specific aspects of business models in the healthcare domain. After that, we outline the research approach and assess alternative business models that refer to the value for providers. This chapter is based on the work by De Vos, Haaker, and Moen (2007). Note that PDD is not a service using mobile technology. However, we experienced that the STOF model and method can also be applied to other types of (electronic) services, especially when multiple parties are involved in the provisioning network, such as the case for PDD.

17.2 About PDD

The Personalized Dementia Directory is an interactive digital service specifically designed for people with dementia and their informal carers. Although it is not a mobile service, it is available through the Internet. In the future the Internet application will also be made available for mobile devices. The purpose of the service is to provide personalized answers to questions of people with dementia and informal carers concerning the care and support they need in their daily lives. PDD helps users focus their questions and subsequently provides answers that consist of relevant services and service bundles that meet their specific needs and that are available in the region where they live. Work on the design and specification of user requirements was carried out by, among others,

Baida, Gordijn, Sale, Akkermans, and Morch (2005) and Dröes et al. (2005). User studies showed that people with dementia and their carers require specific support to help them cope with the consequences of the disease, especially with memory problems and problems in their daily activities. However, variation, fragmentation and continuous changes in care and welfare services make it (more) difficult to access the services. By offering an improved access to optimal care, PDD is expected to improve service usage and with it the quality of life of people with dementia and their carers. Although as yet no specific impact figures are available, it is expected that PDD will reduce the costs per patient-carer dyad. Enabling people with dementia and carers to access and benefit from the services they need, will prolong the period that people with dementia can continue to live in their own homes and delay institutionalization, and it will prevent the carer from becoming overburdened. From a societal point of view, the cost reduction involved is important in the long term, because dementia is a relatively costly disease and the number of people with dementia is expected to increase dramatically in decades to come (Lamura, 2003; Qiu, De Ronchi, & Fratiglioni, 2007).

The core research of PDD focuses on the service and technology domain. The domain descriptions are provided below. Next, some generic issues will be addressed with regard to the organization and finance domain.

17.2.1 Service Domain

The PDD service was developed by researchers in the healthcare domain based on the results of a comprehensive literature study (Van der Roest, Meiland, et al., 2007) and a large scale needs survey among Dutch dementia patients and their carers (Van der Roest, Meiland, et al., 2007). The survey aimed at exploring the various (met and unmet) needs and wishes of the respondents (Dröes et al., 2005). PDD provides personalized information on the disease and on cure, care, support and welfare services that aim at solving specific user problems. It provides access to information and services that otherwise would be hard to find. One of the requirements of the PDD service was that the effort users had to make had to be minimal. In the first iteration of the service domain design, pricing was not an issue. The use of the service is illustrated in Fig. 17.1. Different groups of users (people with dementia, informal carers and health care professionals) can ask their questions (demands) and are given an answer in the form of a single service or a combined set of services. The services are provided by a network of suppliers.

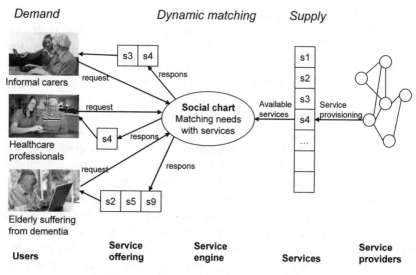

Fig. 17.1. PDD request and response

17.2.2 Technology Domain

The technology domain was specified together with the service design, starting with the main challenge: an engine – algorithm – that could translate user needs into specific demands and match these demands with the available (service) information packages (or bundles) (Baida et al., 2005). A domain expert was involved to specify the information supply and to define the range of needs that should be dealt with by PDD. Next, a user interface was developed and small scale user tests were conducted to adjust the prototype to the specific requirements of users. The global technical PDD architecture is outlined in Fig. 17.2, with the user and research roles and their activities at the top, e.g. informal carers that determine user needs, receive service bundles and determine the context, and several domain experts that classify needs, specify regulations and consequences, and specify specific aspects of the services. Roles and activities are linked to system functionality and information at the bottom of the architecture. System functionalities are e.g. the engine that matches supply and demand, and several administrative functions that support the experts in their specification activities. For an explanation of the Archimate notation, we refer to Lankhorst (2005).

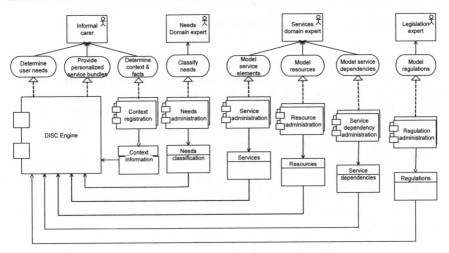

Fig. 17.2. PDD architecture

17.2.3 Issues in PDD's Organization and Finance Domain

PDD's most important output is to offer information on dementia-related cures, care, support and welfare services. This information was specified by domain experts (dementia care researchers advised by care professionals and a board member of the regional Alzheimer Association). Although it is assumed that, in later stages, input and updates are provided directly by the (healthcare) service providers, these providers were not involved in the research project, except for some individual professionals. This presents a potential risk for future exploitation. Another issue is that, during the research project, no organization or institute could be identified that was explicitly interested in exploiting the PDD service. Although the benefits for users are obvious, it is not clear what the benefits would be for such an organization. Individual benefits may add up to a significant societal benefit, since the care of dementia patients is improved, which may in turn reduce the overall costs in caring for these patients. An obvious partner would, therefore, be a government institution; however, the commercial exploitation of these kinds of services is not a governmental task. As a result, the main questions within the organization and finance domains are:

- What are potential value networks for the commercial exploitation of PDD?
- What are the benefits for network partners and how should the costs and benefits be distributed among them?

17.3 Business Model Design Process

Taking the service and technology domain development as a starting point, further research was carried out with respect to designing the organization and finance domains. In addition to exploring potential value networks, one of the aims was to involve stakeholders and identify their specific wishes with regard to the technology and service domain, which would make it possible to produce a balanced business model.

The process that was followed to generate a complete business model is visualized in Fig. 17.3. Via quick-scan sessions with the relevant stakeholders and evaluation sessions concerning the organizational arrangement, we could fine tune the service concept and technology design into a business model.

As a starting point, five alternative business models were constructed, with the aim of promoting discussion with potential stakeholders in the PDD network that might be in a position to provide the service. The initial business models were developed on the basis of a scan of existing business models for online information systems in health-care, additional desk research and discussions with domain experts.

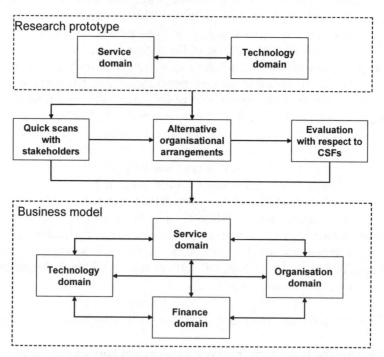

Fig. 17.3. Translation of a service model into business model

The business models were used as input for the discussions with stakeholders, in rounds of interviews and workshops. Since the research was relatively explorative in nature, a limited number of stakeholders were involved.

Stakeholder Interviews. There are numerous stakeholders with regard to PDD, ranging from care and welfare providers to government institutions, insurance companies and interest group, such as informal care communities aimed at supporting informal carers and patient organizations. In principle, all the participants in the study could become part of the future PDD providing network. In all, we interviewed 14 stakeholder representatives, using a semi-structured interview protocol. The questions focused on options for financial and organizational structures for PDD, and addressed issues such as the pros and cons of alternative business models and potential tangible and intangible benefits for the organizations involved.

The first series of interviews with twelve stakeholders provided insight into PDD's expected customer value, as well as into the preferred business models and potential value for providers. An in-depth evaluation of the perceived business model viability was performed in a second series of two workshops and four additional interviews with stakeholder representatives. In this second round, we used the Analytic Hierarchy Process (AHP) introduced by Saaty (1980) to support evaluation of the business model alternatives. As a result, the workshop participants were able to select their optimal business model for PDD. In all, nine people took part in the second round.

Choosing Between Alternative Business Models. After scanning relevant existing literature, we found several approaches that can be used to select business models. Weiss and Amyot (2005), for example, use a goal-oriented requirement language to compare alternatives in terms of their impact on profitability and risk. The authors apply this method to evolving business models. Barabba et al. (2002) use a formal process for decision-makers and supported decisions, combining various approaches from management science, e.g. conjoint analysis, system dynamics models. These approaches require a fairly detailed knowledge regarding several aspects of the business models, which at the time the PDD case was being developed we did not possess. We therefore used a light weight approach by applying the AHP technique to compare alternative business models, and used evaluation criteria derived from the business model success factors of the STOF model.

The AHP technique for multi-criteria decision support was designed by enables users to build a hierarchy with relevant criteria and sub-criteria. Alternatives are evaluated with respect to the goal by pair-wise comparisons. A final judgment is achieved by weighing the subsequent evaluations on the basis of the importance of the criteria, where importance measures are achieved by pair-wise comparisons. In the case of evaluation by multiple participants, average estimates are obtained by arithmetical means. AHP allows users to be inconsistent in their pair-wise comparisons, as indicated by an inconsistency ratio. For an explanation on calculation methods, inconsistency measures and calculation of group results, we refer to Hummel (2001) and Saaty (1980). The AHP method was applied by using the Expert Choice application (*Expert Choice*, 2006).

17.4 Results

During the interviews, all the stakeholders confirmed the value of PDD for people with dementia and their carers. Currently, it is very difficult for these patients and their carers to find the services they need in time, with all consequences, including problems in daily functioning, unsafe situations, accidents, overburdened carers and premature admission into nursing homes. PDD fits into a modern society where people are used to make their own choices. The stakeholders furthermore stated that PDD would be a valuable tool for professionals and healthcare consultants, since they also lack a proper overview of available services. This implies potential new target groups in the service domain. Since PDD helps people find services, it should also be able to reflect which services are missing and where market opportunities arise. However, focusing solely on dementia did not fit with the strategies of the care providers, who offer their services to customers with various diseases and disabilities.

Within the four business model domains, a variety of design issues was addressed. For example, stakeholders stressed the importance of the quality of information (service domain) and provided suggestions with regard to implementation, like asking feedback from users, and striving for a uniform way of presentation. Within the organization domain, issues like maintaining an organization's identity were important, which may conflict with the requirement mentioned above. With regard to the financial aspects of PDD, care providers were reluctant to invest in such a system. They stressed the importance of finding a balance in contribution and returns. Next to a general reflection regarding the opportunities and potential value of PDD, several business models for exploitation were discussed with the stakeholders.

17.4.1 Alternative Business Models and Stakeholders Assessment

A general role model or value network for PDD is visualized in Fig. 17.4. The PDD provider, responsible for providing the PDD service to the users, and for maintaining the service, occupies a central position in this model. Providers of care and welfare services, ranging from taxi services and housekeepers to medical care and cure, are responsible for providing accurate, up-to-date information regarding their services. Information providers are responsible for providing general information, like information on dementia. Advertisers and sponsors may share revenues with the PDD provider in exchange for advertising or marketing opportunities. Finally, there is the user, in most cases the informal carer or a person with dementia.

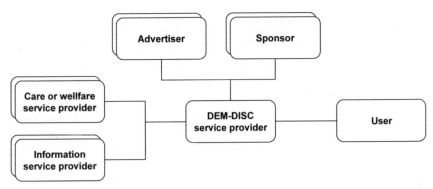

Fig. 17.4. Basic PDD role model

Five organizational arrangements can be derived from the role model, by allocating the role of PDD provider to a specific actor. This allocation affects, among other things, the value proposition and the user group. The five models are:

- *Commercial Model.* PDD is provided to the general public by a commercial party. All providers are allowed to provide information about their services, provided they meet a certain quality standard. Revenues from sponsors and advertisers could be used to cover exploitation costs. Potential sources of income are subscription fees from care providers.
- *Community Model.* A patient community takes on the central role of PDD provider. Care providers that meet specific quality standards are allowed to present their services. The patient community will be an important provider of information on disease and specific care and

support alternatives. Quality of information is an issue that requires specific attention.

- *Government Model.* PDD is provided by a government institution. It is unclear how the quality of information can be guaranteed. The primary aim is to create transparency and a secondary goal could be to enhance competition.
- *Provider Model.* A group of care providers cooperates to provide PDD to the general public. Services to be included are likely those provided by network partners and complementary ones.
- *Insurer Model.* PDD is provided by an insurance company to its own customers. Services provided are likely to be biased and limited to associated providers or providers that are preferred by the insurance company.

Table 17.1 summarizes the (qualitative) opinions on the alternative arrangements as expressed by stakeholders representatives in the two rounds of interviews and workshops, including the main pros and cons.

Table 17.1. Perceived viability of the business models

Arrangement	Pro	Con	Average assessment
Commercial model	There is a direct interest in providing the best possible service, since this will generate revenues for the company	Commercial parties may also have other interests that conflict with that of PDD	Not viable
Community model	Patient communities represent the interests of patients and are independent	There are doubts on the level of professionalism among patients communities	Could be viable
Government model	Governments have a general interest in the wellbeing of elderly people	Governments have no direct interests and focus mainly on short-term policies	Viable
Provider model	Providers have a direct interest in informing customers about their services	Providers are not used to working together with competitors	Could be viable
Insurer model	Insurers have the financial means, power and relevant relations to make it a success	Insurers are dominant parties in the market and pointing to preferred providers is undesirable	Could be viable

Most stakeholders preferred the community model. The government model was also positively evaluated by most parties, except for the government parties themselves. The insurer model and provider model are not considered viable options. There were also serious doubts about the commercial model. Interviewees stated that there is friction between being commercial and providing objective information. The insurer model is not considered a feasible option for PDD exploitation, since dementia care is not covered in health insurance policies, although this might change in the near future due to changes in healthcare regulations.

17.4.2 AHP Business Model Evaluation

To obtain greater insight into potential PDD partners, we conducted interviews and workshops where stakeholders could reflect in detail on each of the business model alternatives. We adopted a structured approach using AHP. Criteria that were important for the viability of business models were derived from the first series of interviews. There is a match between the criteria and the Critical Success Factors (CSFs) as defined in Chap. 3. These are quality of information, i.e. complete, accurate and up to date; acceptable division of profits, i.e. sufficient benefits and an acceptable distribution of costs, benefits and investments; acceptable division of roles, i.e. all roles can be assumed by capable actors and partners should be comfortable with the role division; and clear network strategy, i.e. alignment of visions on the service. Other CSFs are not relevant to this case.

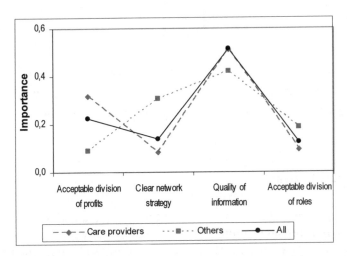

Fig. 17.5. The importance of success factors in business model assessment

Figure 17.5 summarizes the evaluation rates of the decision criteria. To distinguish results between respondents, we divided them into two groups: care providers and others (e.g. government, insurance companies and patient organizations). On average, the care providers indicate that profitability is an important element of a PDD business model. The other participants disagree. According to the care providers, having a clear network strategy is relatively unimportant, whereas it is highly valued by the others. Quality of information is the most important element, even more so to the care providers than to the others. An acceptable division of roles is considered to be less important by all the participants, although there is some variation. We emphasize that the results only give an indication, keeping in mind that not all stakeholders were involved, the small number of interviews and the substantial variation between the individual ratings. Figure 17.6 shows the average evaluation rate with regard to the care providers and the other stakeholders, including the range of the rates. The other stakeholders consider the commercial model the most attractive. The provider model is not preferred by this group. In this regard, their preferences more or less mirrored the evaluation of the care providers, who prefer a business model where the providers are the initiators and are reluctant to join a commercial model. The differences between care providers and others with regard to the community model versus the government model are small.

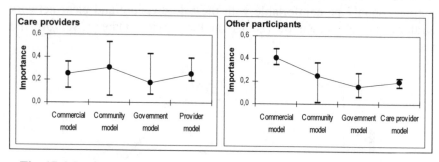

Fig. 17.6. Business model assessments by AHP for care providers and others

17.5 Conclusion

In this chapter we illustrated the use of the STOF method to move from a user-centered research prototype (i.e. mainly service and technology domain) towards a design that involves the needs and requirements of the potential providers. Ideally, the supply side would be included in a research project. However, in practice this is not always easy. In our research we asked potential stakeholders to assess PDD's assumed value for users and providers. Furthermore, potential providers reflected on and evaluated five potential business models and their own role in these models. Issues that they considered important were discussed in relation to all the relevant business model domains. For example, the way quality of information could be guaranteed and how a company's identities could be safeguarded. Such issues are not likely to be addressed in user-related studies. By elaborating on business model alternatives, stakeholders were able to distinguish the pros and cons of the business models.

Due to a large variation between stakeholders' opinions, no specific model emerged that was preferred by all. Some stakeholders strongly base their opinion on return on investments, while others focus on business models of which they assume that a high quality of information provision can be achieved. Although one might argue that the model that on average gained the highest level of popularity should be selected, that would mean that most of the stakeholders would not be satisfied with the result, since the outcome would not reflect their preferred model, and they would not be likely to cooperate. The most feasible option would be to mix the models, for instance the government model and the provider model, taking the resulting combination as a starting point for the exploitation of PDD.

This study has shown that exploring stakeholder opinions is essential for a viable exploitation of systems like PDD. It has yielded additional requirements that would not have emerged in user studies and undisclosed additional opportunities of PDD for providers. Talking about business models turned out to be an excellent way to involve stakeholders, giving them the opportunity to provide professional feedback and focus on the link between PDD and their product range. Although the study has not resulted in a definite business model, we feel that it has brought such a business model, and the associated service, a small step towards market introduction, which is why the discussion with potential stakeholders, such as care providers, the municipality, the government, insurers, patient organizations and commercial parties will be continued in the near future, the ultimate aim being to create a viable business model for the exploitation of PDD that ensures quality of information about care and welfare services for a group of patients in urgent need.

Abbreviations

3G	Third Generation Mobile Telecommunication
3GPP	3rd Generation Partnership Project
ADSL	Asymmetric Digital Subscriber Line
AHP	Analytical Hierarchy Process
ARPU	Average Revenue Per User
ATM	Automatic Teller Machine
B2C	Business to Customer
BPM/BPMN	Business Process Modeling Notation
BREW	Binary Runtime Environment for Wireless
CAPEX	Capital Expenditure
CD	Compact Disc
CDI	Critical Design Issue
CDMA-2000	Code Division Multiple Access-2000
cHTML	Compact Hypertext Markup Language
CPE	Customer Provided Equipment
CRM	Customer Relation Management
CRUD	Create, Read, Update and Delete
CSF	Critical Success Factor
DECT	Digital Enhanced (formerly European) Cordless Telecommunications
DSL	Digital Subscriber Line
ebXML	eBusiness Extended Markup Language
EDGE	Enhanced Data rates for GSM Evolution
FMEA	Failure Mode and Effect Analysis
FOMA	Freedom of Mobile Multimedia Access
FRC	Foreign Remittance Center
GIF	Graphics Interchange Format
GIS	Geographic Information System
GPRS	General Packet Radio Service/System
GPS	Global Positioning System
GSM	Global System for Mobile communications
HSDPA	High-Speed Data Packet Access
HSPA	High-Speed Packet Access
HSUPA	High-Speed Uplink Packet Access

HTML	Hypertext Markup Language
HTTP	Hypertext Transfer Protocol
ICT	Information and Communication Technology
ID	Identity
IDEF	Integration Definition Function Modeling
IETF	Internet Engineering Task Force
IMS	IP Multimedia System
IP	Internet Protocol
IPTV	Internet Protocol Television
ISP(s)	Internet Service Provider(s)
IT	Information Technology
J2ME	JAVA Micro Edition
JAVA	computer software products and specifications
LAN	Local Area Network
LBS	Location Based Services
M2Mapplications	Machine-to-Machine Applications
Mbps	Megabit Per Second
MMS	Multimedia Messaging Service
MPEG-4	Moving Picture Experts Group-4
MSN	Microsoft Network
MVNO	Mobile Virtual Network Operators
MWS	Mobile Web Services
NFC	Near Field Communication
OMA	Open Mobile Alliance
OPEX	Operational Expenditure
P2P	Peer-to-Peer or Person-to-Person
PAN	Personal Area Network
PC	Personal Computer
PDA	Personal Digital Assistant
PDD	Personal Dementia Directory
PESTEL	Political, Economic, Social, Technical, Environment and Legislative
PIM	Presence and Instant Messaging/Personal Information Management
PLC	Product Life Cycle
PVR	Personal Video Recorder
QFD	Quality Function Deployment
R&D	Research and Development
RFID	Radio Frequency Identification
SADT	Structure Analysis and Design Technique
SCM	Supply Chain Management
SIM	Subscriber Information Module

SMS	Short Message Service
SOA	Service Oriented Architecture
SOAP	Simple Object Access Protocol
SSL	Secure Sockets Layer
STOF	Service, Technology, Organization, Finance
TAM	Technology Acceptance Model
TMC	Travel Management Channel
UDDI	Universal Description, Discovery and Integration
UML	Unified Modeling Language
UMTS	Universal Mobile Telecommunications System
USB	Universal Serial Bus
USSD	Unstructured Supplementary Service Data
UTAUT	Unified Theory of Acceptance and Use of Technology
UWB	Ultra Wide Band
VoD	Video on Demand
WAP	Wireless Application Protocol
W-CDMA	Wideband Code Division Multiple Access
WiFi	A Wireless technology
WiMAX	Worldwide Interoperability for Microwave Access
WML	Wireless Mark-up Language
WSDL	Web Services Description Language
WWRF	Wireless World Research Forum
xHTML	Extended Hypertext Markup Language
XML	Extended Markup Language

References

Abel. (2007). *Homepage of Abel*. Retrieved January 7, 2008, from Http://www.Uitmetabel.Nl/

Acquisti, A., & Grossklags, J. (2004). Privacy attitudes and privacy behaviors. In L. Camp & S. Lewis (Eds.), *Economics of information security* (pp. 165–178). Dordrecht: Kluwer

Adams, W., & Yellen, J. (1976). Commodity bundling and the burden of monopoly. *Quarterly Journal of Economics, 90*(3), 475–498

Afuah, A., & Tucci, C. (2001). *Internet business models and strategies*. Boston: McGraw-Hill

Ajit, K., & Van Heck, E. (2002). *Making markets*. Boston: Harvard Business School Press

Ajzen, I. (1991). The theory of planned behavior. *Organizational Behavior and Human Decision Processes, 50*(2), 179–211

Alcatel. (2007). *Overview of worldwide IPTV activities*. Retrieved March 17, 2007, from www.Alcatel.Com/Tripleplay/Graphics/19320_Iptvworld_600.Jpg

Ali Eldin, A. M. T. (2006). *Towards private information sharing under uncertainty: Dynamic consent decision-making mechanisms for context-aware mobile services*. Delft, The Netherlands: Delft University of Technology

Allee, V. (2000a). Reconfiguring the value network. *Journal of Business Strategy, 21*(4), 36–39

Allee, V. (2000b). The value evolution. Addressing larger implications of an intellectual capital and intangibles perspective. *Journal of Intellectual Capital, 1*(1), 17–32

Alt, R., & Zimmerman, H. (2001). Introduction to special section – Business models. *Electronic Markets, 11*(1), 3–9

Amit, R., & Zott, C. (2001). Value creation in e-business. *Strategic Management Journal, 22*(6–7), 493–520

Anckar, B., & D'Incau, D. (2002). Value creation in mobile commerce: Findings from a consumer survey. *Journal of Information Technology and Application, 4*(1), 43–64

Ansoff, H. I. (1987). Strategic management of technology. *IEEE Engineering Management Review, 15*(3), 2–13

Appiah-Adu, K., Fyall, A., & Singh, S. (2001). Marketing effectiveness and business performance in the financial services industry. *Journal of Services Marketing, 15*, 18–34

Arnold, M. (2003). On the phenomenology of technology: The "Janus-Faces" of mobile phones. *Information and Organization, 13*(4), 231–256

AT Kearney. (2003). *Finding new answers to the pricing question: What is it worth to the customer, Illinois, USA*. Retrieved January 9, 2008, from Http://www.Atkearney.Com/Shared_Res/Pdf/Value_Based_Pricing_S.Pdf

AtosOrigin. (2006). *Meeting the billing challenge for the telecom industry.* Retrieved January 24, 2008, from Http://www.Atosorigin.Com/NR/Rdonlyres/ C5BDE4DF-13F6-4EB3-86A0-3519F46DC958/0/Wp_Telco_Billing.Pdf

Axelsson, B., & Easton, G. (Eds.). (1992). *Industrial networks, a new view of reality.* London: Routledge

Baida, Z., Gordijn, J., Sale, H., Akkermans, H., & Morch, A. Z. (2005). *An ontological approach for eliciting and understanding needs in E-services.* Paper presented at the 17th international conference on advanced information systems engineering, Porto, Portugal

Bakos, Y., & Brynjolfsson, E. (1999). Bundling information goods: Pricing, profits, and efficiency. *Management Science, 45*(12), 1613–1630

Bakos, Y., & Brynjolfsson, E. (2000). Bundling and competition on the internet. *Marketing Science, 19*(1), 63–82

Ballon, P. (2004). Scenarios and business models for 4G in Europe. *Info, 6*(6), 363–382

Ballon, P. (2007). Business modelling revisited: The configuration of control and value. *Info, 9*(5), 6–19

Barabba, V., Huber, C., Cooke, F., Pudar, N., Smith, J., & Paich, M. (2002). A multi method approach for creating new business models: The general motors onstar project. *Interfaces, 32*(1), 20–34

Barnes, S. J. (2002). The mobile commerce value chain: Analysis and future developments. *International Journal of Information Management, 22*(2), 91–108

Barney, J. R. (1991). Firm resources and structural competitive advantage. *Journal of Management, 17*(1), 99–120

Barney, J. R. (2001). The resource based view of the firm: Ten years after 1991. *Journal of Management, 27*(6), 625–641

Barras, R. (1986). Towards a theory of innovation in services. *Research Policy, 15*(4), 161–173

Barras, R. (1990). Interactive innovation in financial and business services: The vanguard of the service revolution. *Research Policy, 19*(3), 215–237

Berg, M. (1999). Patient care information systems and healthcare network: A socio-technical approach. *International Journal of Medical Informatics, 55*(2), 87–101

Berger, S., Mcfaddin, S., Narayanaswami, C., & Raghunath, M. (2003). *Web services on mobile devices – Implementation and experience.* Paper presented at the 5th IEEE workshop on mobile computing systems and applications, Monterey, California

Bergholtz, M., Jayaweera, P., Johannesson, P., & Wohed, P. (2003). Process models and business models – A unified framework. In A. Olivé, M. Yoshikawa, & E. Yu (Eds.), *Advanced conceptual modeling techniques* (pp. 364–377). Berlin Heidelberg New York: Springer

Bergman, S., Frissen, V. A. J., & Slaa, P. (1995). Gebruik en betekenis van de telefoon in het leven van alledag (Use and meaning of telephone in daily life). In Toeval of noodzaak: Geschiedenis van overheidsbemoeienis met de informatievoorziening (Co-incidence or necessity: History of government occupation in information provisioning) (pp. 277–327). Den Haag: Rathenau Instituut

Berry, L., Shankar, V., Parish, J., Cadwallader, S., & Dotzel, T. (2006). Creating new markets through service innovation. *MIT Sloan Management Review, Winter*, 56–63

Blazevic, V., Lievens, A., & Klein, E. (2003). Antecedents of project learning and time-to-market during new mobile service development. *International Journal of Service Industry Management, 14*(1), 120–147

Bohlin, E., Lindmark, S., Rodríguez, C., & Burgelman, J.-C. (2006). *Mapping European wireless trends and drivers*. Brussels: European Commission

Booz-Allen, & Hamilton. (1982). *New product management for the 1980s*. New York: Booz-Allen & Hamilton

Borst, P. (1997). *Construction of engineering ontologies for knowledge sharing and re-use*. Enschede, The Netherlands: University of Twente

Boswijk, A., Thijssen, T., & Peelen, E. (2005). *Een nieuwe kijk op de experience economy: Betekenisvolle belevenissen* (A new view on experience economy: Meaningful experiences). Upper Saddle River, NJ: Pearson

Bouwman, H. (2004). *Anywhere, anytime, anyplace or here and now in this context: Third generation mobile services: A Vignet-study*. Paper presented at the 54th annual conference of the ICA

Bouwman, H., & MacInnes, I. (2006). *Dynamic business model framework for value webs*. Paper presented at the 39th annual Hawaii international conference on system sciences, Big Island, Hawaii, January 4, 2006–January 7, 2006

Bouwman, H., & Van Den Ham, E. (2003a). *Exploring value networks enabling the delivery of back office content to mobile workers*. Paper presented at the ITI'03 Europrix conference, Tampere

Bouwman, H., & Van Den Ham, E. (2003b). Business models and emetrics, A state of the art. In B. Preissl, H. Bouwman, & C. Steinfield (Eds.), *Elife after the Dot.com bust* (pp. 609–618). Berlin Heidelberg New York: Springer

Bouwman, H., & Van der Duin, P. (2003). Technological forecasting and scenarios matter: Research into the use of information and communication technology in the home environment in 2010. *Foresight, 5*(4), 8–20

Bouwman, H., Carlsson, C., Molina-Castillo, F., & Walden, P. (2007). Barriers and drivers in the adoption of current and future mobile services in Finland. *Telematics and Informatics, 24*(2), 145–160

Bouwman, H., Carlsson, C., Molina-Castillo, F. J., & Walden, P. (2008). Trends in mobile services in Finland 2004–2006: From ringtones to mobile internet. *Info, 10*(2), 75–93

Bouwman, H., De Vos, H., Faber, E., Haaker, T., Kijl, B., Eldin, A., et al. (2005). *Dynamic provisioning of service bundles: A state of the art*. Enschede, The Netherlands: Freeband, Telematica Instituut

Bouwman, H., Faber, E., & Haaker, T. (2004). *Balancing strategic interests for network value of mobile services.* Paper presented at the 10th AMCIS conference, New York

Bouwman, H., Faber, E., & Haaker, T. (2005). *Designing business models for mobile service bundles.* Paper presented at the Hong Kong mobility roundtable, Hong Kong

Bouwman, H., Faber, E., & Van Der Spek, J. (2005). *Connecting future scenario's to business models of insurance intermediaries.* Paper presented at the 18th Bled ecommerce conference E integrity, Bled, Slovenia

Bouwman, H., Haaker, T., & De Vos, H. (2005). *Designing 3G+business models: A practical approach.* Paper presented at the 26th McMaster world congress, Hamilton, Canada

Bouwman, H., Haaker, T., & De Vos, H. (2007). Mobile service bundles: The example of navigation services. *Electronic Markets, 17*(1), 28–38

Bouwman, H., Haaker, T., Steen, M., & De Vos, H. (2003). *The value proposition from the end user perspective as main driver for creating complex business models.* Paper presented at COST 269, Helsinki, Finland

Bouwman, H., MacInnes, I., & De Reuver, M. (2006). *Dynamic business model framework: A comparative case study analysis.* Paper presented at ITS 2006 – 16th biennial conference, Beijing, China

Bouwman, H., Staal, M., & Steinfield, C. (2001). Klantenervaring en internet concepten. (Consumer experience and internet concepts). *Management & Informatie, 9*(6), 52–60

Bouwman, H., Van De Wijngaert, L., & De Vos, H. (2008). *Context-sensitive mobile services for police officers.* Accepted for: International Conference on M-business. Barcelona, Spain. July 6–8

Bouwman, H., Van Den Hooff, B., Van De Wijngaert, L., & Van Dijk, J. (2005). *Information & communication technology in organizations.* London: Sage

Boyd, C., & Jacob, K. (2007). *Mobile financial services and the under-banked: Opportunities and challenges for M-banking and M-payments.* Chicago, IL: The Center for Financial Service Information

Braet, O., & Ballon, P. (2007). Strategic design issues of IMS versus end-to-end architectures. *Info, 9*(5), 44–56

Brandenburger, A., & Nalebuff, B. (1996). *Co-opetition.* New York: Currency/Doubleday

Broens, T. H. F., Huis In't Veld, R. M. H. A., Vollenbroek-Hutten, M. M. R., Hermens, H. J., Van Halteren, A. T., & Nieuwenhuis, L. J. M. (2007). Determinants of successful tele-medicine implementations: A literature study. *Journal of Tele-Medicine and Telecare, 13*(6), 303–309

Brynjolffson, E., Yu, H., & Smith, M. (2006). From niches to riches. The anatomy of the long tail. *Sloan Management Review, 47*(4), 67–71

Buijs, J. (1984). *Innovation and intervention.* Deventer: Kluwer

Buijs, J., & Valkenburg, R. (1996). *Integrale Productontwikkeling* (Integral product development). Utrecht, The Netherlands: Lemma

Burgelman, R. A. (1983). A process model of internal corporate venturing in the diversified major firm. *Administrative Science Quarterly, 28*(2), 223–244

Camponovo, G., & Pigneur, Y. (2003). *Business model analysis applied to mobile business.* Paper presented at the 5th international conference on enterprise information systems, Angres

Carlson, R. (1996). *The information super highway. Strategic alliances in telecommunications and multimedia.* New York: St Martin Press

Carlsson, C. (2006). Special issue on mobile technology and services. *Electronic Commerce Research and Applications, 5*(3), 189–191

Castells, M. (1996). *The information age: Economy, society and culture. Volume 1 The rise of the network society.* Cambridge MA: Blackwell

Chen, Z., & Dubinsky, A. (2003). A conceptual model of perceived customer value in E-commerce: A preliminary investigation. *Psychology & Marketing, 20*(4), 323–347

Chesbrough, H. (2003). *Open innovation: The new imperative for creating and profiting from technology.* Boston: Harvard Business Press

Chesbrough, H., & Rosenbloom, R. S. (2002). The role of the business model in capturing value from innovation: Evidence from xerox corporation's technology spin-off companies. *Industrial and Corporate Change, 11*(3), 529–555

Chesbrough, H., Vanhaverbeke, W., & West, J. (Eds.). (2006). *Open innovation: Researching a new paradigm.* Oxford, UK: Oxford University Press

Chiasson, G. (1999). Bundle of joy. *Telephony, 237*(23), 42–46

Chlamtac, I., Conti, M., & Liu, J. J.-N. (2003). Mobile ad hoc networking: Imperatives and challenges. *Ad Hoc Networks, 1*(1), 13–64

Christensen, C. M. (1997). *The innovator's dilemma. When new technologies cause great firms to fail.* Boston: Harvard Business School

Clark, G., Johnston, R., & Shulver, M. (2000). Exploiting the service concept for service design and development. In J. Fitzsimmons & M. Fitzsimmons (Eds.), *New service develop* (pp. 71–91). Thousand Oaks, CA: Sage

Clausing, D. P. (1994). *Total quality development: A step-by-step guide to world-class concurrent engineering.* New York: ASME Press

Cristiano, J., Liker, J. K., & White, C. C., III. (2000). Customer driven product development through quality function deployment in the US and Japan. *Journal of Product Innovation Management, 17*(4), 286–308

Cross, N. (1994). *Engineering design method: Strategies for product design.* Chichester: Wiley

Cuevas, A., Moreno, J. I., Vidales, P., & Einsiedler, H. (2006). The IMS service platform: A solution for next-generation network operators to be more than bit pipes. *IEEE Communications Magazine, 44*(8), 75–81

Curtis, B., Kellner, M., & Over, J. (1992). Process modeling. *Communications of the ACM, 35*(9), 75–90

Cushnie, J., Hutchison, D., & Oliver, H. (2000). *Evolution of charging and billing models for GSM and future mobile internet services.* Berlin Heidelberg New York: Springer

Dahlberg, T., Mallat, N., Ondrus, J., & Zmijewska, A. (2007). Past, present and future of mobile payments research: A literature review. *Electronic Commerce Research and Applications* (in press)

Das, T. G., & Teng, B. S. (1998). Resources and risk management in the strategic alliance making process. *Journal of Management, 24*(1), 21–42

Davenport, T. (1993). *Process innovation: Reengineering work through information technology*. Boston: Harvard Business School Press

Davis, F. D. (1989). Perceived usefulness, perceived ease of use, and user acceptance of information technology. *MIS Quarterly, 13*(3), 319–340

Deitel, H. M., Deitel, P. J., & Nieto, T. (2001). *e-Business & e-commerce. How to program*. Upper Saddle River, NJ: Prentice-Hall

Deitel, H. M., Deitel, P. J., & Steinbuhler, K. (2001). *e-Business & e-commerce for managers*. Upper Saddle River, NJ: Prentice-Hall

Delaere, S., & Ballon, P. (2007). The business model impact of flexible spectrum management and cognitive networks. *Info, 9*(5), 57–69

Demkes, R. (1999). *COMET: A comprehensive methodology for supporting telematics investment decisions*. Enschede: Telematica Instituut

Den Hertog, P. (2000). Knowledge-intensive business services as co-producers of innovation. *International Journal of Innovation Management, 4*(4), 491–528

De Reuver, M., Bouwman, H., & MacInnes, I. (2007). *What drives business model dynamics? A case survey*. Paper presented at the 8th world congress on the managment of ebusiness, Toronto, Ontario, Canada

De Reuver, M., Haaker, T., & Bouwman, H. (2007). *Business model dynamics: A longitudinal, cross-sectional case survey*. Paper presented at the 20th Bled econference, Bled, Slovenia

De Vos, H., Haaker, T., Kleijnen, M., & Teerling, M. (2008). *Context-aware and location based mobile services and consumers' perceived value*. Paper submitted to ECIS 2008

De Vos, H., Haaker, T. I., & Moen, S. (2007). How to move from research prototype to exploitation phase: The DEM-DISC case. In P. Cunningham & M. Cunningham (Eds.), *Expanding the knowledge economy* (Vol. 4, pp. 615–622). Amsterdam: IOS Press

Devlin, B., Gray, J., Laing, B., & Spix, G. (1999). *Scalability terminology: Farms, clones, partitions, and packs: RACS and RAPS*. Retrieved February 5, 2008, from Http://Arxiv.Org/Abs/Cs.AR/9912010

Dietz, J. L. G. (1996). *Introductie to DEMO: Van Informatietechnologie Naar Organisatietechnologie (Introduction to DEMO: From information technology to organization technology)*. Alphen Aan De Rijn, The Netherlands: Samsom

Doherty, D., Glapa, M., Kamat, S., Magee, F., Prakash, S., & Ruffolo, D. (2004). *Balancing network and business planning for cable telephony*. Paper presented at the 11th international telecommunications network strategy and planning symposium, Vienna, Austria

Droës, R. M., Meiland, F. J. M., Doruff, S., Varodi, I., Akkermans, H., Baida, Z., et al. (2005). A dynamic interactive social chart in dementia care: Attuning demand and supply in the care for persons with dementia and their carers. In L. Bos, S. Laxminarayan, & A. Marsh (Eds.), *Medical and care compunetics 2* (pp. 210–220). Amsterdam: IOS Press

Drop, R., Dijkhuis, J., Van Der Duin, P. A., & Stavleu, H. (2000). *Bestemming 2005: Corporate scenario's voor KPN (Destination 2005: Corporate scenarios for KPN)*. KPN Studieblad, April/May, 204–221

Dym, C., Agogino, A., Eris, O., Frey, D., & Leifer, L. (2005). Engineering design thinking, teaching, and learning. *Journal of Engineering Education, 94*(1), 103–110

ECLAC. (2003). *Road maps towards an information society in Latin America and the Caribbean*. Santiago, Chile: ECLAC

Edvardsson, B., Gustafsson, A., & Enquist, B. (2006). Success factors in new service development and value creation through services. In D. Spath & K.-P. Fähnrich (Eds.), *Advances in service innovations* (pp. 165–189). Berlin Heidelberg New York: Springer

Estenfeld, K. (2006). *Mobile business evolution*. Paper presented at the CICT conference, Copenhagen, Denmark

European Commission. (2003). *The competitiveness of business-related services and their contribution to the performance of European enterprises*. Retrieved January 23, 2008, from Http://Ec.Europa.Eu/Internal_Market/Services/Brs/Competitiveness_En.Htm

European Commission. (2005). *European monitoring centre on change*. Retrieved August, 2006, from Http://Eurofound.Europa.Eu/Emcc/Publications/2005/Ef0567en.Pdf

Evans, P., & Wurster, T. (1999). *Blown to bits: How the economics of information transformsstrategy*. Boston: Harvard Business School Press

Expert Choice. (2006). *Expert choice 11*. Retrieved January, 2008, from Http://www.Expertchoice.Com/Software/

Faber, E., & Bouwman, H. (2003). *Designing business models for mobile services. Exploring the connections between customer value and value networks*. Paper presented at the ICEB 2003, Singapore

Faber, E., Ballon, P., Bouwman, H., Haaker, T., Rietkerk, O., & Steen, M. (2003). *Designing business models for mobile ICT services*. Paper presented at the 16th Bled electronic commerce conference e-transformation, Bled, Slovenia

Faber, E., Haaker, T., Bouwman, H., & Rietkerk, O. (2003). *Business models for personalised real-time traffic information in cars: Which route to take*. Paper presented at the ICEC, Pittsburg

Fähnrich, K.-P., & Meiren, T. (2006). Service engineering: State of the art and future trends. In D. Spath & K.-P. Fähnrich (Eds.), *Advances in service innovations* (pp. 3–16). Berlin Heidelberg New York: Springer

Farley, P., & Capp, M. (2005). Mobile web services. *BT Technology Journal, 23*(2), 202–213

Feijóo, C., Marín, A. Á., Martín, Á., & Rojo, D. (2006). *European regulation, competition models and mobile content access*. Paper presented at the CICT conference, Copenhagen, Denmark

Fijnvandraat, M., & Bouwman, H. (2006). Flexibility and broadband evolution. *Telecommunication Policy, 30*(8–9), 424–444

Fishbein, M., & Ajzen, I. (1975). *Beliefs, attitude, intention and behaviour: An introduction to theory and research.* Reading, MA: Addison-Wesley

Forrester. (2006a). *The European mobile landscape 2006*

Forrester. (2006b). *Towards an open mobile internet (Presentation for open mobile internet initiative)*

Foster, R., Daymon, C. M., & Tewungwa, S. (2002). *Future reflections. Four scenarios for television in 2012.* Retrieved November, 2007, from Http://Media.Bournemouth.Ac.Uk/Research/Documents/Fullreport.Pdf

Frodigh, M., Parkvall, S., Roobol, C., Johansson, P., & Larsson, P. (2001). Future-generation wireless networks. *IEEE Personal Communications, 8*(5)(October), 10–17

Galbreath, J. (2002). Twenty-first century management rules: The management of relationships as intangible assets. *Management Decision, 40*(2), 116–126

Gallouj, F., & Weinstein, O. (1997). Innovation in services. *Research Policy, 26*(4–5), 537–556

Giaglis, G. M. (2001). A taxonomy of business process modeling and information systems modeling techniques. *The International Journal of Flexible Manufacturing Systems, 13*(2), 209–228

Goldstein, S. M., Johnston, R., Duffy, J. A., & Rao, J. (2002). The service concept: The missing link in service design research? *Journal of Operations Management, 20*(2), 121–134

Gordijn, J., & Akkermans, H. (2001). E3-value: Design and evaluation of E-business models. *IEEE Intelligent Systems, 16*(4), 11–17

Gordijn, J., Akkermans, H., & Van Vliet, H. (2000). Business modelling is not process modelling. In S. W. Liddle & H. C. Mayr (Eds.), *Conceptual modeling for E-business and the web* (pp. 40–51). *LNCS 1921.* Berlin Heidelberg New York: Springer

Grace, D., & O'Cass, A. (2005). Service branding: Consumer verdicts on service brands. *Journal of Retailing and Consumer Services, 12*(2), 125–139

Granovetter, M. (1994). Business groups. In N. J. Smelser & R. Swedberg (Eds.), *The handbook of economic sociology* (pp. 453–475). Princeton: Princeton University Press

Grant, R. M. (1991). The resource based theory of competitive advantage. Implications for strategy formulation. *California Management Review, 33*(3), 114–135

Grönroos, C. (1992). *Service management and marketing: Managing the moment of truth in service competition.* Lexington: Lexington books

Grönroos, C. (1994). From marketing mix to relationship marketing: Towards a paradigm shift in marketing. *Management Decision, 32*(2), 4–20

Grönroos, C. (2000). *Service management and marketing: A customer relation management approach.* New York: Wiley

Grönroos, C. (2007). Service management and marketing: Customer management in service competition (3rd Ed.). Chichester: Wiley

Grönroos, C., Heinonen, F., Isoniemi, K., & Lindholm, M. (2000). The netoffer model: A case example from the virtual marketspace. *Management Decisions, 38*(4), 243–252

Grover, V., & Saeed, K. (2003). The telecommunication industry revisited – The changing pattern of partnerships. *Communications of the ACM, 46*(7), 119–125

Gruber, T. R. (1995). Towards principles for the design of ontologies used for knowledge sharing. *International Journal of Human Computer Studies, 43*(5–6), 907–928

GSMA/IFC. *Micro-payment systems and their application to mobile networks.* Retrieved February 5, 2008, from www.Infodev.Org/En/Publication.43.Html

Guiltinan, J. P. (1987). The price bundling of services: A normative framework. *Journal of Marketing, 51*(2), 74–85

Gulati, R., Nohria, N., & Zaheer, A. (2000). Strategic networks. *Strategic Management Journal, 21*(3), 203–216

Haaker, T., & De Vos, H. (2007). *Customer preferences for bundled content services.* Paper presented at the 18th European regional ITS conference, Istanbul, Turkey

Haaker, T., Bouwman, H., Kijl, B., Galli, L., Killström, U., Immonen, O., et al. (2007). *Challenges in business models for context aware services.* Paper presented at the Collecter Iberoamérica, Cordoba, Argentina

Haaker, T., Faber, E., & Bouwman, H. (2006). Balancing customer and network value in business models for mobile services. *International Journal of Mobile Communications, 4*(6), 645–661

Haaker, T., Galli, L., Kijl, B., Killström, U., Immonen, O., & De Reuver, M. (2006). *Challenges in designing viable business models for context-aware mobile services.* Paper presented at the 3rd international CICT conference, Copenhagen, Denmark

Haaker, T., Oerlemans, K., Steen, M., & De Vos, H. (2004). *STOF business blueprint method. Handbook for successful cooperation in the development and exploitation of innovative (mobile) ICT-services.* Enschede, The Netherlands: Telematica Instituut

Hammer, M. (1990). Reengineering ork: Don't automate, obliterate. *Harvard Business Review, July–August,* 104–112

Harlam, B. A., Krishna, A., Lehmann, D. R., & Mela, C. (1995). Impact of bundle type, price framing and familiarity on purchase intention for the bundle. *Journal of Business Research, 33*(1), 57–66

Hauser, J., & Clausing, D. (1988). The house of quality. *Harvard Business Review, 66*(3), 63–73

Hawkins, R. (2002). The phantom of the marketplace: Searching for new E-commerce business models. *Communication & Strategies, 46*(2), 297–329

Hedman, J., & Kalling, T. (2003). The business model concept: Theoretical underpinnings and empirical illustrations. *European Journal of Information Systems, 12*(1), 49–59

Hegering, G. G., Küpper, A., Linnhoff-Popien, C., & Reiser, H. (2004). *Management challenges of context-aware services in ubiquitous environments.* Paper presented at the DSOM, Heidelberg, Germany

Hekkert, M. P., Suurs, R. A. A., Negro, S. O., Kuhlmann, S., & Smits, R. E. H. M. (2007). Functions of innovation systems: A new approach for analyzing technological change. *Technological Forecasting & Social Change, 74*(4), 413–432

Henderson, J. C., & Venkatraman, N. (1993). Strategic alignment: Leveraging information technology for transforming organisations. *IBM Systems Journal, 32*(1), 4–16

Herzwurm, G., Schockert, S., Dowie, U., & Breidung, M. (2002). *Requirements engineering for mobile-commerce applications.* Paper presented at the M-commerce conference, Athens

Heskett, J. L., Sasser, W. E., & Schlesinger, L. A. (1997). *The service profit chain: How leading companies link profit and growth to loyalty, service and value.* New York: Free Press

Hill, C., & Jones, G. (1995). *Strategic management: An integrated approach.* Boston: Houghton Mifflin

Hofacker, C. F., Goldsmith, R. E., Bridges, E., & Swilley, E. (2007). E-Services: A synthesis and research agenda. *Journal of Value Chain Management, 1*(1/2), 14–44

Holland, Chr., Bouwman, H., & Smidts, M. (2001). *Return to the bottom line.* Leidschendam: ECP NL

Howells, J. (2006). *Role of R&D services in service innovation.* Manchester: RENESER

Hughes, J., Lang, K. R., & Vragov, R. (2007). An analytical framework for evaluating peer-to-peer business models. *Electronic Commerce and Applications, 7*(1), 105–118

Huis In't Veld, R. M. H. A., Fielt, E., Faber, E., & Vollenbroek-Hutten, M. M. R. (2007). *Business model for mobile tele-health services: Phase 1: Developing the business model.* Enschede, The Netherlands: Freeband, Telematica Instituut

Hummel, J. M. (2001). *Supporting medical technology development with the analytic hierarchy process.* Groningen, The Netherlands: Rijksuniversiteit Groningen

Idenburg, P. J. (2005). *Oog voor de Toekomst. Over Marketing en Consumenten in een Veranderende Samenleving (Eye on the future about marketing and consumers in a changing society).* Schiedam, The Netherlands: Scriptum Management

Ilie, V., Van Slyke, C., Green, G., & Lou, H. (2005). Gender differences in perceptions and use of communication technologies: A diffusion of innovation approach. *Information Processing Management Journal, 18*(3), 13–31

Immonen, O., Haaker, T., Galli, L., Killström, U., Pitkänen, O., & De Reuver, M. (2006). *Can advertising based earnings logic become a basis for future mobile models?* Paper presented at the 3rd International CICT conference, Copenhagen, Denmark

Jaokar, A., & Fish, T. (2004). *Open gardens: The innovator's guide to the mobile data industry.* London, UK: Futuretext

Johnson, G., & Scholes, K. (2002). *Exploring corporate strategy – Text and cases* (6th ed.). Harlow, Essex: Prentice-Hall

Johnson, S. P., Menor, L. J., Roth, A. V., & Chase, R. B. (2000). A critical valuation of the new service development process. In J. Fitzsimmons & M. Fitzsimmons (Eds.), *New service development* (pp. 1–32). Thousand Oaks, CA: Sage

Johnston, R. (1999). Service operations management: Return to roots. *International Journal of Operations & Production Management, 19*(2), 104–124

Jonason, A., & Holma, B. (2002). *Pricing for profits on the mobile internet.* Paper presented at the IEEE international engineering management conference

Jones, C., Hesterly, W., & Borgatti, S. (1997). A general theory of network governance: Exchange conditions and social mechanisms. *The Academy of Management Review, 22*(4), 911–945

Kaasinen, E. (2005). User acceptance of location-aware mobile guides based on seven field studies. *Behaviour and Information Technology, 24*(1), 37–49

Kambil, A., & Short, J. (1994). Electronic integration and business network redesign: A roles-linkage perspective. *Journal of Management Information Systems, 10*(4), 59–83

Kaplan, R., & Norton, D. (1992). The balanced scorecard: Measures that drive performance. *Harvard Business Review, January–February*, 71–79

Kaplan, R., & Norton, D. (1996). Using the balanced scorecard as a strategic management system. *Harvard Business Review, January–February*, 75–85

Katz, M. L. (2002). Industry structure and competition absent distribution bottlenecks. In E. Noam, J. Groebel, & D. Gerbarg (Eds.), *Internet television.* Mahwah, NJ: Lawrence Erlbaum Associates

Kiiski, A. (2006). *Impacts of MVNOs on mobile data service market.* Paper presented at the regional ITS conference, Amsterdam, The Netherlands

Kijl, B., & Timmerman, W. (2003). *Presence and instant messaging: Cross case analysis.* Enschede: Telematica Instituut

Killström, U., Virola, H., Galli, L., Immonen, O., Pitkänen, O., & Kijl, B. (2006). *Business models for new mobile applications and services.* Retrieved February 27, 2008, from Http://www.ist-mobilife.org/ (D1.5)

Killström, U., Galli, L., Haaker, T., Immonen, O., & De Reuver (2007). Marketplace dynamics and business models framework. In M. Klemettinen (Ed.), *Enabling technologies for mobile services: The mobilife book.* Chichester, UK: Wiley

Klein, S., & Loebbecke, C. (2003). Emerging pricing strategies on the web: Lessons from the airline industry. *Electronic Markets, 13*(1), 46–58

Klein-Woolthuis, R. (1999). *Sleeping with the enemy, trust and dependence in inter organisational relationships.* Enschede, The Netherlands: University of Twente

Klemettinen, M. (Ed.). (2007). *Enabling technologies for mobile services: The mobilife book.* Chichester, UK: Wiley

Konczal, E. F. (1975). Models are for managers, not mathematicians. *Journal of Systems Management, 26*(1), 12–15

Koolwaaij, J., Tarlano, A., Luther, M., Morhs, B., Battestini, A., & Vaidya, R. (2006). *Contextwatcher – Sharing context information in everyday life.* Paper presented at the IASTED international conference on web technologies, applications, and services (WTAS2006), Calgary, Canada

Koppenjan, J., & Groenewegen, J. (2005). Institutional design for complex technological systems. *International Journal of Technology, 5*(3), 240–257

Kothandaraman, P., & Wilson, D. (2001). The future of competition. Value creating networks. *Industrial Marketing Management, 30*(4), 379–389

Kotler, P. (1988). *Marketing management: Analysis, planning, implementation, and control* (6th ed.), Englewood Cliffs, NJ: Prentice-Hall

Kotler, P. (1999). *Principles of marketing.* London: Prentice-Hall

Kotler, P. (2000). *Marketing management – International Edition – The Millennium Edition.* Englewood Cliffs, NJ: Prentice-Hall

Kotler, P., Armstrong, G., Saunders, J., & Wong, V. (1996). *Principles of marketing: The European edition.* Hemel Hempstead, UK: Prentice-Hall

Koushik, S., & Joodi, P. (2000). E-Business architecture design issues. *IT Pro, May/June*, 38–43

Koutsopoulou, M., Kaloxylos, A., & Alonistioti, A. (2004). Charging, accounting and billing management schemes in mobile telecommunication networks and the internet. *IEEE Communications Surveys, 6*(1), 50–58

Krueger, M. (2001). *The future of M-payments, business options and policy issues.* Retrieved February 5, 2008, from Http://Fiste.Jrc.Es/Pages/Detail.Cfm?Prs=750

Kubr, T., Marchesi, H., Ilar, D., & Kienhuis, H. (1998). *Starting up – Achieving success with professional business planning.* Zürich, Switzerland: Mckinsey

Kumar, V. (2001). Wireless communications "Beyond 3G". *Alcatel Telecommunications Review, 1*(1), 28–32

Kuo, Y.-F., & Yu, C.-W. (2006). 3G telecommunication operators' challenges and roles: A perspective of mobile commerce value chain. *Technovation, 26*, 1347–1356

Kurvinen, E. (Ed.). (2006). *Results of service and application evaluation.* Retrieved January 28, 2008, from Http://www.Ist-Mobilife.Org, (D1.9)

Lamura, G. (2003). *Supporting carers of older people in Europe: A comparative report on six European countries.* Paper presented at the 11th European social services conference, Venice

Lankhorst, M. (Ed.). (2005). *Enterprise architecture at work.* Berlin Heidelberg New York: Springer

Lankhorst, M., Van Der Stappen, P., & Jansen, W. (2001). *State of the art in E-business services and components.* Enschede: Telematica Instituut

Larsson, R. (1993). Case survey methodology: Quantitative analysis of patterns across case studies. *Academy of Management Review, 39*(6), 1515–1546

Law, A., & Kelton, D. (1982). *Simulation modeling and analysis.* New York: McGraw-Hill

Lee, C.-S. (2001). An analytical framework for evaluating E-commerce business models and strategies. *Internet Research: Electronic Networking Applications and Policy, 11*(4), 349–359

Leem, C. S., Suh, H. W., & Kim, D. W. (2004). A classification of mobile business models and its applications. *Industrial Management & Data Systems, 104*(1), 78–87

Li, F., & Whalley, J. (2002). Deconstruction of the telecommunications industry: From value chains to value networks. *Telecommunications Policy, 26*(9–10), 451–472

Limbu, D. K., Wah, L. E., & Yushi, C. (2004). Wireless web services clients development – Using web services standards and J2ME technology. *Synthesis Journal, 5,* 151–162

Limonard, S. (2006). *User driven business models for digital television. Exploring the long tail for audiovisual content on TV and the internet.* Enschede, The Netherlands: Freeband, Telematica Instituut

Limonard, S., & Teem, R. (2007). *User driven business models – Assessing the value of the long tail for audiovisual services.* Retrieved January 31, 2008, from Http://www.Cs.Tut.Fi/~Lartur/Euroitv07_Ajp/WIP7.Htm

Liu, N. (2006). *Business model, operation model & profit model.* Retrieved March 18, 2007, from Http://www.Huawei.Com/Publications/View.Do?Id=685& Cid=342&Pid=61

Loebbecke, C., & Radtke, S. (2005). *Business models and programming choice: Digital video recorders shaping the TV industry.* Paper presented at the 11th Americas conference on information system, Omaha

López-Nicolás, C., Molina-Castillo, F. J., & Bouwman, H. (Draft). *An assessment of advanced mobile services acceptance: Contributions from TAM and diffusion theory models.* Delft, The Netherlands: Delft University of Technology

Loucopoulos, P., & Karakostas, V. (1995). *Systems requirement engineering.* New York: Mcgraw- Hill

Lovelock, C. H. (1984). Developing and implementing new services. In W. R. George & C. E. Marshall (Eds.), *Developing new services* (pp. 44–64). Chicago: American Marketing Association

Low, J., & Cohen Kalafut, P. (2002). *Invisible advantage. How intangibles are driving business performance.* Cambridge: Persues Publishing

Luczak, H., Gill, C., & Sander, B. (2006). Architecture for service engineering. The design and development of industrial service work. In D. Spath & K.-P. Fähnich (Eds.), *Advances in service innovations* (47–63). Berlin Heidelberg New York: Springer

Magretta, J. (2002). Why business models matters. *Harvard Business Review, May,* 86–92

Mahadevan, B. (2000). Business models for internet – Based E-commerce. *California Management Review, 42*(4), 55–69

Maitland, C. F., Bauer, J. M., & Westerveld, R. (2002). The European market for mobile data: Evolving value chains and industry structures. *Telecommunications Policy, 26*(9–10), 485–504

Maitland, C., Van De Kar, E., Wehn De Montalvo, U., & Bouwman, H. (2005). Mobile Information and Entertainment Services: Business Models and Service Networks. *International Journal of Management and Decision Making, 6*(1), 47–64

Mallat, N., Rossi, M., & Tuunainen, V. K. (2004). Mobile banking services. *Communications of the ACM, 47*(5), 42–46

Malone, T. W. (1987). Modeling coordination in organizations and markets. *Management Science, 33*(10), 1317–1332

Markus, M. L. (1990). Toward a critical mass theory of interactive media. In J. Fulk & C. W. Steinfield (Eds.), *Organizations and communication technology* (pp. 194–218). Newbury Park, CA: Sage

Mason, H., & Rohner, T. (2002). *The venturing imperative*. Boston: Harvard Business Press

Matthing, J., Kristensson, P., Gustafsson, A., & Parasuraman, A. (2006). Developing succcesfull technology-bases services: The issue of identifying and involving innovative users. *Journal of Service Marketing, 20*(5), 288–297

Maxmilien, E., & Sigh, M. (2004). A framework and ontology for dynamic web services selection. *Internet Computing, 8*(5), 84–93

Mcnaughton, R., Osborne, P., & Imrie, B. (2002). Market-oriented value creation in service firms. *European Journal of Marketing, 36*(9/10), 990–1002

Meiren, Th. (2006). *Services' R&D as performed by manufacturing firms*. Stuttgart: Springer

Melão, N., & Pidd, M. (2000). A conceptual framework for understanding business processes and business process modelling. *Information Systems Journal, 10*(2), 105–129

Menor, L. J., Tatikonda, M. V., & Sampson, S. E. (2002). New service development: Areas for exploitation and exploration. *Journal of Operations Management, 20*(2), 135–157

Methlie, L. B., & Pedersen, P. E. (2007). Business model choices for value creation of mobile services. *Info, 9*(5), 70–86

Microsoft. (2003). *Mobile web services roadmap*. Retrieved January 24, 2008, from Http://www.Microsoft.Com/Serviceproviders/Resources/Bizresmwsroadmap. Mspx

Miles, I., Keenan, M., & Kaivo-Oja, J. (2002). *Handbook of knowledge foresight society*. Manchester: PREST/FFRC

Millen, D. R. (2000). *Rapid ethnography: Time deepening strategies for HCI field research*. Paper presented at the 3rd conference on designing interactive systems: Processes, practices, methods, and techniques, New York, NY

Miller, D., & Shamsie, J. (1996). The resource based view of the firm in two environments. The hollywood firm studios from 1836 to 1964. *Academy of Management Journal, 39*(3), 519–543

Miller, R., & Lessard, D. (2000). *The strategic management of large engineering projects. Shaping institutions, risks and governance*. Boston: MIT Press

Mintzberg, H. (1983). *Structure in fives: Designing effective organizations*. Englewood Cliffs, NJ: Prentice-Hall

Mobilife. (2008). *Homepage of mobilife*. Retrieved January 28, 2008, from www.Ist-Mobilife.Org

Mobinet. (2002). Retrieved in 2005, from Http://www.Mobile.Commerce.Net/Story.Php?Story_Id=1471&S=2. Link is no longer available

Mogg, A. (2004). *Limits of free-TV (presentation)*. Cologne: University of Cologne

Monge, P., & Contractor, N. (2003). *Theories of communication networks.* Oxford: Oxford University Press

Moon, Y. (2005). Break free from the product life cycle. *Harvard Business Review, 83*(5), 86–94, 153

Mooney, J. G., Gurbaxani, V., & Kreamer, J. L. (1995). *A process oriented framework for assessing the business value of information technology.* Paper presented at the 16th annual international conference on information systems, Amsterdam, The Netherlands

Morris, M., Schindehutte, M., & Allen, J. (2005). The entrepreneur's business model: Towards a unified perspective. *Journal of Business Research, 58*(6), 726–735

Myers, J. (2002). *Consumer interest in DVR-features.* Retrieved June, 2004, from www.Jackmyers.Com/Pdf/03-27-02.Pdf

Nalebuff, B. (2004). Bundling is an entry barrier. *Quarterly Journal of Economics, 119*(1), 159–187

Neely, A., Adams, C., & Kennerley, M. (2002). *The performance prism. The scorecard for measuring and managing business success.* London: Prentice Hall

Netsize. (2007). *The Netsize guide.* Retrieved January 24, 2008, from Http://www.Netsize.Com/Downloads/Registration.Aspx

Niemegeers, I. G., & Heemstra De Groot, S. M. (2003). Research issues in ad-hoc distributed personal networks. *Wireless Personal Communications, 26*(2–3), 149–167

Noam, E. (2002). Will internet TV be American? *Trends in Communication, 11*(2), 101–109

Nokia, & Sun. (2004). *Deploying mobile web services using Liberty Alliance's Identity Web Services Framework (ID-WSF).* Retrieved January 24, 2008, from Http://www.Sun.Com/Software/Products/Identity/Sun_Nokia_Id-Wsf_Deployment_Paper.Pdf

Normann, R. A. (2000). *Service management: Strategy and leadership in service business.* Chichester: Wiley

Oliver, P., Marwell, G., & Texeira, R. (1985). A theory of critical mass I. Interdependence, group heterogeneity, and the production of collective action. *American Journal of Sociology, 91*(3), 522–556

Olla, P., & Patel, N. V. (2002). A value chain model for mobile data service providers. *Telecommunications Policy, 26*(9–10) 551–571

O'Mahony, M., & Van Ark, B. (2003). *EU productivity and competitiveness: An industry perspective. Can Europe resume the catching-up process?* Brussels: European Commission

OMG. (2006). *Business process modeling notation Specification.* Retrieved February 11, 2008, from Http://www.Bpmn.Org

Ondrus, J., & Pigneur, Y. (2005). *A disruption analysis in the mobile payment market.* Paper presented at the 38th annual Hawaii international conference on system sciences, Hawaii

Ondrus, J., & Pigneur, Y. (2006). Towards a holistic analysis of mobile payments: A multiple perspectives approach. *Electronic Commerce Research and Applications, 5*(3), 246–257

O'Neill, P., & Sohal, A. (1999). Business process reengineering: A review of recent literature. *Technovation, 19*(9), 571–581

Osterwalder, A., & Pigneur, Y. (2002). *An e-business model ontology for modelling e-business.* Paper presented at the 15th Beld electronic commerce conference, Bled, Slovenia

OVUM. (2006). *Off-portal developments and operator strategy.* London: OVUM

Panagiotakis, S., Koutsopoulou, M., & Alonistioti, A. (2005). *Business models and revenue streams in 3G market.* Paper presented at the 14th IST mobile and wireless communications, Summit, Dresden, Germany

Parasuraman, A., Zeithaml, V., & Berry, L. (1988). SERVQUAL: A multiple item scale for measuring consumer perceptions of service quality. *Journal of Retailing, 64*(1), 168–174

Pashtan, A. (2005). *Mobile web services.* Cambridge, UK: Cambridge University Press

Petrovic, O., & Kittl, C. (2003). *Capturing the value proposition of a product or service.* Paper presented at the international workshop on business models, Lausanne, Switzerland

Pfeffer, J., & Salancik, G. (1978). *The external control of organizations. A resource dependence perspective.* New York: Harper & Row

Pigneur, Y., (2004). An ontology for M-business models. *Lecture Notes in Computer Science, 2503/2002,* 3–6

Pilioura, T., Tsalgatidou, A., & Hadjiefthymiades, S. (2003). Scenarios of using web services in M-commerce. *ACM Sigecom Exchanges, 3*(4), 28–36

Pine, J., & Gilmore, J. (1999). *The experience economy.* Boston: Harvard Business School Press

Porter, M. (1980). *Competitive strategy.* New York: The Free Press

Porter, M. (1985). *Competitive advantage: Creating and sustaining superior performance.* New York: The Free Press

Porter, M. (1990). *The competitive advantage of nations.* London: The Macmillan Press

Porter, M. (2001). Strategy and the internet. *Harvard Business Review, March,* 63–76

Porter, M., & Millar, V. (1985). How information gives you competitive advantage. *Harvard Business Review, 63*(4), 149–160

Powell, M. W. (1990). Neither market nor hierarchy: Network forms of organisation. *Research in organisational behaviour, 12,* 295–336

Power, A. (2000). Channel surfing. *Outlook, January,* 54–61

Qiu, C., De Ronchi, D., & Fratiglioni, L. (2007). The epidemiology of dementia: An update. *Current Opinion in Psychiatry, 20*(4), 380–385

Raessens, B. (2001). *e-Business your business. Over de effectiviteit van E-commerce* (E-Business your business. About the effectiveness of e-commerce). Utrecht, The Netherlands: Lemma

Räisänen, V., Karasti, O., Steglich, S., Mrohs, B., Räck, C., Del Rosso, C., et al. (2005). *Basic reference model for service provisioning and general guidelines*. Helsinki: Mobilife

Ramaswamy, R. (1996). *Design and management of service processes*. Reading: Addison-Wesley

Rappa, M. (2000). *Business models on the web*. Retrieved February 23, 2006, from Http://Digitalenterprise.Org/Models/Models.Html

Rappa, M. (2001). *Managing the digital enterprise – Business models on the web*. Retrieved October, 2007, from Http://Digitalenterprise.Org/Models/ Models.Html

Rayport, F. (1999). The truth about business internet business models. *Strategy and Business, 16*(3rd Quarter), 1–3

Rayport, J. F., & Jaworski, B. J. (2001). *e-Commerce*. New York: McGraw-Hill

Rayport, J. F., & Sviokla, J. (1994). Managing in the market space. *Harvard Business Review, 72*(6), 141–145

Rayport, J. F., & Sviokla, J. (1995). Exploring the virtual value chain. *Harvard Business Review, 73*(6), 75–85

Recker, J., Indulska, M., Rosemann, M., & Green, P. (2005). *Do process modelling techniques get better? A comparative ontological analysis of BPMN*. Paper presented at the 16th Australian conference on information systems, Sydney, Australia

Rendón, J., Kuhlmann, F., & Aranis, J. P. (2007). *A business case for the deployment of a 4G wireless heterogeneous network in Spain*. Paper presented at the 18th European international telecommunications society conference (ITS), Istanbul, Turkey

Reneser. (2006). *Research and development needs of business related service firms*. Utrecht: Dialogic

Renkema, T. (1996). *Investeren in de Informatie-Infrastructuur. Richtlijnen voor besluitvorming in organisaties (Investing in information infrastructure. Guidelines for decision making in organisations)*. Deventer, The Netherlands: Kluwer Bedrijfsinformatie

Rietkerk, O., & Timmerman, W. (Eds.). (2003). *Communities of interest: Cross case analysis*. Enschede, The Netherlands: Telematica Instituut

Ritter, T., Wilkinson, I., & Johnston, W. (2002). Measuring network competence: Some international evidence. *Journal of Business & Industrial Marketing, 17*(2–3), 119–138

Rockart, J., & Bullen, C., (1981). *A primer on critical success factors*. Cambridge, MA: Sloan School of Management, MIT

Rogers, E. M. (1995). *Diffusion of innovations*. New York: Free Press

Rojas, E. (2005). *The GSM evolution and the mobile revolution in Latin America. Business briefings wireless technology*. Retrieved February 5, 2008, from Http://www.Touchbriefings.Com/Pdf/1433/Rojas.Pdf

Roland Berger. (2005). *What's the score on mobile data services?* Amsterdam: Roland Berger

Roussel, A., Daum, A., Flint, D., & Riseley, M. (2000). *B2C web business models: Winners and losers*. Gartner Group Research

Rowley, J. (2006). An analysis of the E-service literature: Towards a research agenda. *Internet Research, 16*(3), 339–359

Saaty, T. (1980). *The AHP. Planning, priority setting, resource allocation.* New York: McGraw-Hill

Sabat, H. K. (2002). The evolving mobile wireless value chain and market structure. *Telecommunications Policy, 26*(9–10), 505–535

Sakaguchi, T., Nicovich, S., & Dibrell, C. (2004). Empirical evaluation of an integrated supply chain model for small and medium sized firms. *Information Resources Management Journal, 17*(3), 1–19

Salter, A., & Tether, B. (2006). *Innovation in services: Through the looking glass of innovation studies.* A background review paper prepared for the inaugural meeting of the Grand Challenges in Services (GCS) forum, held at Said Business School, Oxford, May 2006

Sandy, J.-C., & Bouwman, H. (2006). *Un Marco Conceptual Para El Análisis Y Diseño De Modelos De Negocio De Servicios Moviles En Latinoamérica (Designing mobile services: Conditions for the future development of mobile services in Latin America).* Paper presented at the 12th Americas conference on information systems, Acapulco, Mexico

Schoenbachler, D. D., & Gordon, G. L. (2002). Multi-channel shopping: Understanding what drives channel choice. *Journal of Consumer Marketing, 19*(1), 42–53

Schuler, D., & Namioka, A. (1993). *Participatory design: Principles and practices.* Hillsdale, NJ: Lawrence Erlbaum Associates

Schumpeter, J. (1934). *Theory of economic development.* Oxford: Oxford University Press

Scott Morton, M. (Ed.). (1990). *The corporation of the 1990s: Information technology and organisational transformation.* New York: Oxford University Press

Selz, D. (1999). *Value webs. Emerging forms of fluid and flexible organizations.* Retrieved January 26, 2008, from Http://En.Scientificcommons.Org/12748919

Shafer, S. M., Smith, H. J., & Linder, J. C. (2005). The power of business models. *Business Horizons, 48*(4), 199–207

Shapiro, C., & Varian, H. (1999). *Information rules: A strategic guide to the network economy.* Boston, MA: Harvard Business School Press

Shostack, L. G. (1984). Designing services that deliver. *Harvard Business Review, 62*(1), 133–139

Silverstone, R., & Haddon, L. (1996). Design and the domestication of information and communication technologies: Technical change and everyday life. In R. Mansell & R. Silverstone (Eds.), *Communication by design. The politics of information and communication technology* (pp. 44–74). Oxford: Oxford University Press

Silverstone, R., & Hirsch, E. (Eds.) (1992). *Consuming technologies: media and information in domestic spaces.* London: Routledge

Simonin, B. L., & Ruth, J. A. (1995). Bundling as a strategy for new product introduction: Effects on consumers' reservation prices for the bundle, the new product, and its tie. *Journal of Business Research, 33*(3), 219–230

Simons, L. (2006). *Multi-channel services for click and mortars.* Delft, The Netherlands: Delft University of Technology

Simons, L., & Bouwman, H. (2004). Developing a multi channel service model and design method. *International Journal of Internet Marketing and Advertisement, 1*(3), 229–250

Simons, L., & Bouwman, H. (2005). Multi channel service design processes: Challenges and solutions. *International Journal of Electronic Business, 3*(1), 50–67

Smith, A., Mitra, S., & Narasimhan, S. (1998). Information systems outsourcing: A study of pre-event firm characteristics. *Journal of Management Information Systems, 15*(2), 61–93

Stabell, C. B., & Fjeldstad, Ø. D. (1998). Configuring value for competitive advantage: On chains, shops, and networks. *Strategic Management Journal, 19*(5), 413–437

Stähler, P. (2001). *Gesellschäftsmodelle in Der Digitalen Ökonomie. Merkmale, Strategien Und Auswirkungen (Business models for the digital economy. Branding, strategies and consequences).* St Gallen: University of St Gallen

Steen, M., Van Eijk, R., Gunther, H., Hooreman, S., & De Koning, N. (2005). *We-centric services for police officers and informal carers.* Paper presented at the HCI close to you: Sigchi.Nl conference, The Hague, The Netherlands

Strauss, B., Schmidt, M., & Schoeler, A. (2005). Customer frustration in loyalty programs. *International Journal of Service Industry Management, 16*(3), 229–252

Stremersch, S., & Tellis, G. J. (2002). Strategic bundling of products and prices: A new synthesis for marketing. *Journal of Marketing, 66*(1), 55–72

Tafazolli, R. (Ed.). (2005). *Technologies for the wireless future.* London: Wiley

Tanriverdi, H., & Iacano, C. S. (1998). *Knowledge barriers to diffusion of tele-medicine.* Paper presented at the international conference on information systems, Helsinki, Finland

Tanriverdi, H., & Iacono, C. S. (1999). Diffusion of tele-medicine: A knowledge barrier perspective. *Telemedicine Journal, 5*(3), 223–244

Tapscott, D., Lowi, A., & Ticoll, D. (2000). *Digital capital – Harnessing the power of business webs.* Boston: Harvard Business School Press

Tax, S. S., & Stuart, F. I. (1997). Designing and implementing new services: The challenges of integrating service systems, *Journal of Retailing, 73*(1), 105–134

Taylor, S., & Todd, P. (1995). Decomposition and crossover effects in the theory of planned behavior: A study of consumer adoption. *International Journal of Research in Marketing, 12*(2), 137–155

Teerling, M., Haaker, T., & De Vos, H. (2007). *Understanding consumer value of mobile service bundles.* Paper presented at the networking and electronic commerce research (NAEC 2007), Lake Garda, Italy

Teng, J., & Kettinger, W. (1995). Business process redesign and information architecture: Exploring the relationships. *Database Advances, 26*(1), 30–42

Tian, M., Voigt, T., Naumowicz, T., Ritter, H., & Schiller, J. (2004). Performance considerations for mobile web services. *Computer Communications, 27*(11), 1097–1105

Tidd, J., Bessant, J., & Pavitt, K. (2001). *Managing innovation.* New York: Wiley

Tilson, D., & Lyytinen, K. (2006). The 3G transition: Changes in the US wireless industry. *Telecommunications Policy, 30*(10–11), 569–586

Timmers, P. (1998). Business models for E-commerce. *Electronic Markets, 8*(2), 3–7

TNO. (2006). *Marktrapportage Elektronische Communicatie December 2006.* Delft, The Netherlands: TNO

Turban, E., Lee, J., King, D., & Chung, H. M. (2000). *Electronic commerce: A managerial perspective.* Upper Saddle River, NJ: Prentice-Hall

Turpeinen, P. (1998). *Tarpeet Ja Motiivit. Uusmedia Kuluttajan Silmin (Needs and motives. new media through the eyes of a consumer).* Helsinki, Finland: Tekes

UMTS Forum. (2002). *Report 21 charging, billing and payment views on 3G business models.* London: UMTS Forum

UMTS Forum. (2006). *3G/UMTS evolution: Towards a new generation of broadband mobile services.* London: UMTS Forum

Van Bossuyt, M., & Van Hove, L. (2007). Mobile payment models and their implications for NextGen Msps. *Info, 9*(5), 31–43

Van De Kar, E. A. M. (2004). *Designing mobile information services: An approach for organisations in a value network.* Delft, The Netherlands: Delft University of Technology

Van De Kar, E. A. M., & Verbraeck, A. (2007). *Designing mobile service systems.* Delft, The Netherlands: IOS Press

Van Der Duin, P. A (2000). Destination 2005. *Scenario & Strategy Planning, 2*(1), 22–25

Van Der Heijden, K. (1996). *Scenarios. The art of strategic conversation.* Chichester: Wiley

Van Der Roest, H. G., Meiland, F. J. M., Comijs, H. C., Vernooij-Dassen, M. J. F. J., Van Hout, H., Jonker, C., et al. (2007). *What do you need? That's the question. The subjective and objective needs of people with dementia living in the community.* Paper presented at the Alzheimer Europe congress, Estoril, Portugal

Van Der Roest, H. G., Meiland, F. J. M., Maroccini, R., Comijs, H. C., Jonker, C., & Dröes, R. M. (2007). Subjective needs in people with dementia. A review of the literature. *International Psychogeriatrics, 19*(3), 559–592

Van Oirsouw, R., Spaanderman, J., & De Vries, H. (1993). *Informatie-economie: investeringsstrategie voor de informatievoorziening (Information economy: Investment strategy for data processing).* Schoonhoven, The Netherlands: Academic Service

Van Weering, M., Jones, V., & Thiele, F. (2007). *Chronic pain application: Trial, platform and results description.* Enschede: Freeband

Vargo, S. L., & Lusch, R. F. (2004a). The Four service marketing myths. *Journal of Service Research, 6*(4), 324–335

Vargo, S. L., & Lusch, R. F. (2004b). Evolving to a new dominant logic for marketing. *Journal of Marketing, 68*(1), 1–17

Varian, H. R. (1996). Differential pricing and efficiency. *First Monday, 1*(2), Retrieved December 19, 2004, from Http://www.Firstmonday.Dk/Issues/Issue2/Different/Index.Html

Venkatesh, R., & Kamakura, W. (2003). Optimal bundling and pricing under a monopoly: Contrasting complements and substitutes from independently valued products. *The Journal of Business, 76*(2), 211–231

Venkatesh, V., Morris, M. G., Davis, G. B., & Davis, F. D. (2003). User acceptance of information technology: Toward a unified view. *MIS Quarterly, 27*(3), 425–478

Verschuren, P., & Hartog, R. (2005). Evaluation in design-oriented research. *Quality & Quantity, 39*(6), 733–762

Versteeg, G., & Bouwman, H. (2006). Business architecture: A new paradigm to relate business strategy to ICT. *Information Systems Frontier, 8*(2), 91–102

Veryzer, R. W. (1998). Discontinuous innovation and the new product development process. *Journal of Product Innovation Management, 15*(4), 304–321

Vogel, H. L. (2004). *Entertainment industry economics – A guide for financial analysis* (6th ed.). Cambridge, UK: Cambridge University Press

VWS. (2004). *Healthcare in an aging society: A challenge for all European countries*. The Hague, The Netherlands: Ministry of Health, Welfare and Sports

Wang, Y. (2008). *A studio based approach for business engineering and mobile services*. Delft, The Netherlands: Delft University of Technology

Ward, J., & Daniel, E. (2005). *Benefits management: Delivering value from IS & IT investments*. Hoboken, NJ: Wiley

Weill, P., & Vitale, M. R. (2001). *Place to space. Migrating to e-business models*. Boston: Harvard Business School Press

Weiss, M., & Amyot, D. (2005). *Design and evolution of E-business models*. Paper presented at the 7th IEEE international conference on E-commerce Technology, CEC

White, S. (2004). *Process modelling notations and workflow patterns*. Retrieved February 22, 2008, from Http://www.Omg.Org/Bp-Corner/Bp-Files/Process_Modeling_Notations.Pdf

Wigand, R., Picot, A., & Reichswald, R. (1997). *Information, organization and management*. New York: Wiley

Wohltorf, J., & Albayrak, S. (2005). *Decision Cockpit for Mobile Services*. Retrieved January 5, 2008, from Http://www.Dai-Labor.De/Fileadmin/Files/Publications/Wohltorf-Albayrak_2003_BPMJ_Decision-Cockpit-For-Mobile%20Services.Pdf

World Resources Institute. *What works: Smart communications – Expanding networks, expanding profits. Providing telecommunications services to low-income markets in the Philippines*. Retrieved February 5, 2008, from Http://www.Digitaldividend.Org/Pdf/Smart_Communications_Case.Pdf

Yin, R. K., & Heald, K. A. (1975). Using the case survey method to analyze policy studies. *Administrative Science Quarterly, 20*(3), 371–381

Zmijewska, A. (2007). *Factors influencing adoption and diffusion of mobile payments systems. A holistic framework.* Sydney: University of Technology

Authors

Harry Bouwman, Ph.D., is associate professor at and Interim Chair of the Information and Communication Technology section, Faculty of Technology, Policy and Management of Delft University of Technology, the Netherlands, and private docent at Institute for Advanced Management System Research, Åbo Akademi, Turku, Finland.

Tim De Koning, M.Sc., received his Master degree from Faculty of Technology, Policy and Management, Delft University of Technology. At the moment he is employed by KPMG IT Advisory, The Hague, the Netherlands.

Mark De Reuver, M.Sc., received his Master degree from Delft University of Technology, and is now Ph.D. candidate at the Information and Communication Technology section of Delft University of Technology, the Netherlands.

Henny De Vos, Ph.D., is scientific researcher at the Networked Business group of Telematica Instituut, Enschede, the Netherlands.

Rose-Marie Dröes, Ph.D., is associate professor in Psychogeriatrics and program leader of the research group Care and support in Dementia at the Department of Psychiatry, the Alzheimer Centre and the Institute for Research in Extramural Medicine of the VU University Medical Centre in Amsterdam.

Edward Faber, Ph.D., is scientific researcher at the Networked Business group of Telematica Instituut, Enschede, the Netherlands.

Erwin Fielt, Ph.D., is scientific researcher at the Networked Business group of Telematica Instituut, Enschede, the Netherlands.

Ralph Feenstra, M.Sc., received his Master degree from Delft University of Technology, and is now Ph.D. candidate at the Information and Communication Technology section of Delft University of Technology, the Netherlands.

Timber Haaker, Ph.D., is scientific researcher at the Networked Business group of Telematica Instituut, Enschede, the Netherlands.

Rianne Huis In 't Veld, Ph.D., is a researcher at Roessingh Research and Development, Enschede, the Netherlands.

Björn Kijl, M.Sc., received his Master degree from the faculty of Electrical Engineering, Mathematics and Computer Science, and is now Ph.D. candidate at the School of Management and Governance University of Twente, Enschede, the Netherlands.

Sander Limonard, M.A., is a researcher and consultant at TNO Information and Communication Studies, Department of Strategy, Policy and Innovation, Delft, the Netherlands.

Zhengjia Meng, M.Sc., received his Master degree from Faculty of Technology, Policy and Management, Delft University of Technology, the Netherlands. At the moment he is employed by Shell.

Jean-Carlo Sandy, M.Sc., is a Ph.D. candidate at the Information and Communication Technology section, Delft University of Technology, the Netherlands.

Hidde Schipper, M.Sc., received his Master degree from Faculty of Technology, Policy and Management, Delft University of Technology, the Netherlands. Currently he is working as a business analyst at Royal Philips Electronics, Boston.

Marc Steen, M.Sc., received his Master degree in Industrial Design Engineering at Delft University of Technology, and is currently working at TNO Information and Communication Technology. His Ph.D. research is on human-centered design.

Patrick Van Der Duin, Ph.D., is assistant professor at the Strategy, Technology & Entrepreneurship section, Faculty of Technology, Policy and Management, Delft University of Technology. He is specialized in futures research and innovation management.

Miriam Vollenbroek-Hutten, Ph.D., is manager of the research cluster Technology Assisted Pain Rehabilitation at Roessingh Research and Development, Enschede, the Netherlands.

Index